The Patrick Moore Practical Astronomy Series

More information about this series at http://www.springer.com/series/3192

First Light and Beyond

Making a Success of Astronomical Observing

D. A. Jenkins

D. A. Jenkins
Spring, TX, USA

ISSN 1431-9756   ISSN 2197-6562   (electronic)
The Patrick Moore Practical Astronomy Series
ISBN 978-3-319-18850-8   ISBN 978-3-319-18851-5   (eBook)
DOI 10.1007/978-3-319-18851-5

Library of Congress Control Number: 2015052629

Springer Cham Heidelberg New York Dordrecht London

Cover illustration: The Trifid Nebula, courtesy of ESO

Printed on acid-free paper

Springer International Publishing AG Switzerland is part of Springer Science+Business Media
(www.springer.com)

Preface

You own a telescope because you want to see the universe with your own eyes. Anyone can look up pictures of the universe on the Internet, but you want to personally connect with these beautiful celestial treasures and gain a deeper understanding of what you see. There is something very special about visual observing. Real-time observations at the eyepiece can give us such an ethereal, yet visceral feeling that connects us with our universe in very unique ways. After acquiring their telescope, many have wondered, "Now what?" Time for first light! First light is the moment you turn your telescope optics toward the heavens and take in that first celestial sight through the eyepiece. But then after this moment more questions begin to arise. How can I make the best use of my telescope? What can I see with it? How can I get the most out of my instrument? This book will show the beginner how to make a success of astronomical observing.

Beginners today are privileged to live in a time when so many great professional and amateur astronomers have already paved the way for us and provided a fantastic context within which to begin a journey into visual astronomy. There are many texts available for either those who know absolutely nothing about astronomy or for the experienced who are ready to find very challenging objects with little or no assistance. One of the goals here is to fill in the gap between those two stages of development.

In this text we will assume that you have already acquired your first telescope, although if you have not yet done so you will still find information here to be extremely helpful. Amateur astronomy has come a long way over the last 30 years in terms of availability and lower cost; therefore most readers will no doubt have at least a 6-in. (152 mm) scope or even larger. Although many objects covered in this book will have that aperture in mind, many of the objects will still be visible in a

4-in. scope. This book is not intended to provide exhaustive coverage of the various types of telescopes and how they function. On the other hand this book is not simply a database of hard to find objects. Instead what you will find here is a bridge for the newcomer between both of these areas, as progress is made toward becoming a seasoned observer. You will find that this book is easy to understand if you are a beginner, providing step-by-step guidance on how to navigate your way throughout the constellations to many deep sky objects. In view of this, certain concepts or suggestions are repeated throughout the book. There is a wise saying that says, "repetition is the mother of retention." This should help you to remember these important points and to put them within the current context they are used in, as this book helps you to progressively build on those key ideas.

Further, this book will teach you how to make a success of each time you go out under the night sky, regardless of what you planned to observe. It will help you to develop and maintain enthusiasm for observing celestial wonders. You will gain a better understanding of what constitutes successful observing and be more motivated to get out under the stars to create your own memorable visual observing experiences. Several examples are provided in Chaps. 4, 5, and 6 of how to find objects that you will want to see. Star charts are included with most of the featured objects to help you see exactly how to find them. Some charts have notations or drawings that have been added by the author to make the path to an object more apparent. These stellar treks will show you how to hunt down deep sky objects by developing your ability to recognize star patterns (asterisms) such as the Big Dipper, beginning with the easiest steps. This book also uncovers the nature of the objects you are viewing, and how best to locate, recall, and record your observations. It also provides planning tips on making a trip to dark skies, public observatories, and much more.

First Light and Beyond—Making a Success of Astronomical Observing is not simply a collection of facts. It is a journey of amateur astronomy, from the backyard all the way to where the road meets dark rural skies. In this book are practical suggestions presented by someone who has a passion for observing, amid a background of what it is really like to observe with modest amateur telescopes. You do not have to own a computer targeting system or CCD camera to enjoy this book. You will see how you can have successful observing sessions, using equipment and tools that you already own, or that are easy to obtain.

Spring, TX   D. A. Jenkins
March 2015

Acknowledgements

After completing this rewarding project, I am moved to express my immense gratitude to several people. First, this book is dedicated to my father, who fed my passion for astronomy by giving me my first telescope. And to my mother, who is the most positive person I know. And to my wife for her loving support of this project. And to my brother, whose creative genius continues to inspire me. It is also dedicated in memory of Ingrid Jonsson, my middle school librarian, who always set aside the latest issue of *Sky & Telescope* magazine for me.

Special thanks goes to Kathie Coil, the Public Affairs Sr. Program Coordinator for the NOAO Office of Education and Public Outreach, and to the National Optical Astronomy Observatory for permitting use of their outstanding images. I am also very appreciative of the insight received from both Dr. Walker (Associate Scientist & Senior Education Specialist at the NOAO) and John Goss, the current president of the Astronomical League. Special thanks also goes to Greg Crinklaw of Skyhound, the creator of *Sky Tools 3*, for granting the use of his astronomy software to create the charts used in this publication.

I am also extremely grateful to John Watson for being the first to believe in my vision of this book and for his ongoing encouragement. I also appreciate and am grateful for the patience and support of Senior Editor Maury Solomon, Assistant Editor Nora Rawn, and to the entire Production Team at Springer Publishing that had a role in making this book's publication possible.

Contents

About the Author

D. A. Jenkins has had a passion for writing since his teen years, when he resolved to produce a body of work that would both satisfy the needs of readers and his passion to create. Since then he has written business material for several large institutions, organizations, and individual clients, and has also created various short fictional works. He studied financial planning at the University of St. Thomas in Houston, Texas, and has a broad array of professional experience that includes being a classical piano teacher. However, it is his enthusiasm for the night sky that has continued to fuel his passion for astronomy across four decades.

Jenkins has been an enthusiastic amateur astronomer since childhood, with a special interest in deep sky observing. He resides in Texas, where he enjoys observing celestial phenomena in the pristine dark skies of the Davis Mountains, along with many other locations. He thoroughly enjoys astronomy outreach activities such as sharing stunning telescopic sights with people at star parties. Jenkins is a member of the Austin Astronomical Society (www.austinastro.org) and also observes regularly with the North Houston Astronomy Club. He welcomes your emailed comments about this publication at contact@dajenkins.net.

Chapter 1

Planetary Discovery and the Seven Sisters, or How We Fell in Love with the Universe

For each of us, the story of how we fell in love with the universe is different. So what is your story? What caused you to fall in love with the universe? Perhaps it was seeing the cloud bands of Jupiter in a telescope for the first time, or catching sight of the blue stars of the Pleiades as they sparkled like blue diamonds. Or maybe it was discovering the nebulous patch beneath Orion's Belt, gazing up at Venus the "morning star" and realizing that its light shines as a flat disk rather than twinkling, or simply being fascinated with the shapes that stars appear to form from our point of view here on Earth. Whatever it was, hold on to that first fascination and passion for the stars. Now is the time to remember and feed this passion so that it will grow. Your desire to go beyond first light is the initial step that will allow you to make a success of visual astronomical observing (Fig. 1.1).

Visual astronomy is such an exciting and accessible branch of study. To enjoy it you do not have to travel to another country, but at the same time it allows you to journey to far away worlds. One of its beauties is that you can simply walk outside, look up, and find yourself tapping into a science that has been followed by people all over the world for thousands of years. When we look up and see a shooting star, or the colors of the aurora borealis dancing in the sky, the feelings we have are part of a universal language that to a degree is generally understood by everyone. This excitement can be felt even when you observe alone, with only the stars as your companions, or it can be shared with a friend or a group of people who seek the same enjoyment. The amount of celestial objects that can be viewed is endless, so there will never be any shortage of things for you to observe. Even the most experienced amateur observers have not come close to seeing everything there is to view through amateur telescopes. Take a moment to consider that the *New General Catalog* (published in 1888 by Johann Louis Emil Dreyer) contains more than 7,800

© Springer International Publishing Switzerland 2015
D. A. Jenkins, *First Light and Beyond*, The Patrick Moore Practical Astronomy Series,
DOI 10.1007/978-3-319-18851-5_1

Fig. 1.1 Jupiter as seen by NASA's Cassini spacecraft. The image is in true color, close to how it would appear to the human eye. (Image courtesy of NASA/JPL/Space Science Institute)

celestial objects. Just to be generous let's say you were able to observe 65 of those objects *every month*, it would still take you 10 years to finish the list! By that time, you would probably be ready to enjoy a second look at the ones you started with so many years ago, and that's just one catalog. There are many more, enough to provide any amateur astronomer with a modest-sized telescope a lifetime of celestial wonders to observe!

The light we see today began its journey to us thousands of years ago, and in many cases millions of years, yet visual astronomy remains a precise study. In other words, we can accurately predict when objects will be visible in the sky. For example, if you look up at midnight and see the Pleiades straight overhead in mid-November, exactly 30 days later it will be in the same place but 2 h earlier instead of at midnight. This is because Earth's rotation and orbital movement around the Sun causes objects to rise 4 min earlier each night.

Although it is true that all celestial objects are moving, the stars are so distant from us that in essence they appear stationary to us, and so we can reference where each object can be located by its position or coordinates. Only extremely subtle changes in position occur over several decades, and major movement can take thousands of years for us to perceive. The immediate changes we can perceive are daily, monthly, and seasonally, which can be measured by the precise timing of our rotation around Earth's axis and our orbit around the Sun. This example of precision in astronomy is just the beginning.

Has your telescope seen first light yet? Some readers who have recently acquired one may not yet have taken a look, perhaps being a little timid or having nervous excitement because of wanting to get everything just right or at least read more about visual astronomy. Whenever you do decide to enjoy those first photons shining through the telescope and into your pupil, you are sure to be reminded of why you fell in love with the universe in the first place. In the first third of the book we will make an extensive examination of essential concepts and tools that will be of great help to you in observing. In the next several chapters you will learn how to successfully find celestial objects and recall them later. We will also take a thorough look at the nature of the things we observe, to gain a better appreciation for what we are viewing. What is the distance of these objects from Earth? Some will be rather close to us astronomically speaking, like the star Sirius, which happens to be the brightest star in the night sky and is only about 8.8 light years away.

On the other hand, if you are using at least a 4- to 6-in. telescope (102–152 mm) you will be able to see as far away as quasar 3C 273, which is approximately 2 billion light years away from us! In today's digital age photography has come a long way, so in Chap. 7 we'll explore how to get started in astrophotography. Getting the most out of a trip to an observatory is the topic of Chap. 6. The final three chapters will give advice on how to navigate the sea of astronomical information found in print and on the Internet, and will then conclude with suggestions on how to have enjoyable shared experiences in astronomy and also how to encourage our children to be excited about astronomy. Not only will you learn to be a success in visual observing but you will become well rounded in your grasp of the overall pursuit of amateur astronomy. With patience and persistence, you will eventually arrive at a level of experience that you might not have thought possible before.

The question now is, how do you get there? The answer in short is to learn to master the basics. This chapter will reinforce key concepts that you may already be aware of, but also reveal and fill in any gaps in your present understanding. So even if you already understand these ideas, a review may be beneficial.

We can divide the basics into four main categories:

- Distance, direction, and motion
- Star patterns
- How objects will appear in your telescope
- Proper setup and maintenance of your telescope

Just as a person can become lost while traveling on land, the same thing can happen when viewing objects in the night sky. One important factor in getting around the sky easily is understanding distances—both physical distance and angular distances. Another key to successful navigation through the heavens is to learn the direction in which things are, and the directional relationship between two points. Learning this at the outset will save you a lot of time and possible frustration later on. The good news is that we still use the directions of north, south, east, and west. What you will need to grasp is *celestial* north, east, south, and west. Next, the apparent motion of the stars will be explained. Then we'll turn our attention to getting around in the constellations, along with a discussion of what all of the Greek letters on star maps mean, and other ways of referring to specific stars.

As you may have already experienced, the view through a telescope is usually different from the naked-eye view, so we will need to gain an understanding of how objects will appear under magnification as well as how they are oriented in terms of celestial direction, and how to calculate the field of view for a specific eyepiece. These concepts are crucial in order to be able to steer successfully to the deep sky wonders of your choice. We will conclude this chapter with advice for setting up your telescope with its mount, and aligning it with the finderscope or illuminated non-magnifying finder. We will now take a look at each topic, and along the way we'll be sure to weave in some interesting celestial sights for you to observe that will serve to illustrate various points.

Distance, Direction, and Motion

Distance

As you may know, when it comes to astronomy, distances are given in very large units. As children we are normally first taught in school about the astronomical unit, which is the measurement of Earth's distance (1 AU) from the Sun, or about 92,955,807 miles (149,597,870 km). That unit of measurement is fine for discussing objects within our Solar System, which includes the eight major planets and hundreds of other bodies such as the dwarf planets, moons, asteroids, comets, and meteoroids. Mars is 1.52 AU from the Sun, Jupiter is 5.2 AU, Neptune is 30 AU, and Pluto is 39.5 AU. It is currently thought that the entire Solar System extends no farther out than the reach of the heliosphere (the region wherein the Sun's solar wind exerts influence) at a distance of 100 AU. To put this in perspective, *Voyager 1* and *2*, which were both launched in 1977, have gone beyond the 100 AU mark and, as of this writing, are 128 AU and 105 AU from the Sun, respectively (Fig. 1.2).

When it comes to objects beyond our Solar System, we must use a better system of measurement for distance. Earlier we stated that the hot white star Sirius is about 8.8 light years distant from us. A light year is the distance that light travels in 1 year. Light moves at a rate of 186,282 miles per second (299,792 km). That equates to a distance of about 6,000,000,000,000 (6 trillion) miles per year (9,500,000,000,000 km)! Professional astronomers more often use the term *parsec* (pc), which is 3.26 light years. Thus we would say that Sirius is located at a distance of about 2.7 pc from Earth (8.8 light years ÷ 3.26 = 2.7 pc). If we could travel in a spacecraft at the speed of light, it would take 8.8 years to reach Sirius. In some references you might see the term kiloparsec (1,000 pc) or megaparsec (1 million pc). In this book, however, only astronomical units, light years, or parsecs will be used to describe large astronomical distances.

Since we must deal with such vast distances throughout space, traditional forms of measurement here on Earth will not suffice for use when referring to objects in the night sky. For example, the use of terrestrial forms of measurement such as 10 ft to the left, or 3 m to the right would essentially have no universal meaning. A meter

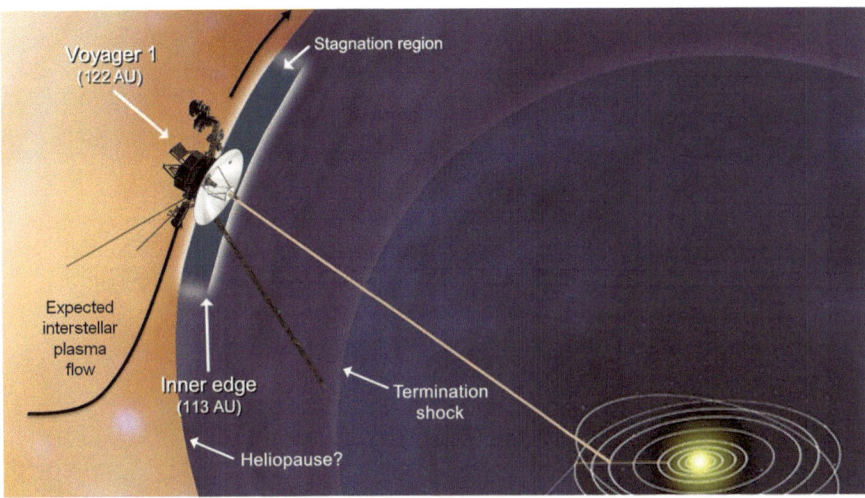

Fig. 1.2 The approximate location of *Voyager 1* in 2012 can be seen as illustrated here, in relation to the Solar System and the Sun's heliosphere. (Image courtesy of NASA/JPL-Caltech/ JHUAPL)

"up" from the star Altair from one person's point of view would be quite different from another person's viewpoint. Instead we measure distance in degrees and then specify the celestial direction (celestial north, east, south, or west) (Fig. 1.3).

To begin grasping this concept let's take a look at the Big Dipper in Ursa Major. Go outside and locate this familiar shape in the northern part of the night sky. Make a fist and extend your hand out to arm's length—the distance from thumb to little finger is about 10°. Compare this estimate with the distance between the two stars on Fig. 1.3 in the bowl of the Dipper labeled Alpha (α) and Beta (β), which is about 5°. The two stars labeled Delta (δ) and Gamma (γ), are about 4° apart. If you stretch out your entire hand, the length of your hand span would be about 20–25° (depending on your hand size), which is about the distance from the star Eta (η), to Alpha (α). Try holding up the three fingers that are in the middle of your hand and they will measure about 5°. Figure 1.4 illustrates some of these points. You can use these as estimates to get around the sky when reading about distances expressed in degrees, and understanding this is very important to being a successful observer.

Another aspect of distance in astronomy relates to a system of measurement that we use to describe and locate objects as they appear on the celestial sphere, and in sky atlases that are used to refer to the positions of these objects. We know that in reality space is not in the shape of a sphere; however, this metaphor helps us to understand the place of objects in relation to Earth. The shape of Earth and its rotation gives us the illusion of this spherical shape. As you know, we use a similar method when locating our position on Earth itself, using lines of latitude and longitude. To understand where objects are located in the sky we must perceive Earth as if it were inside this great celestial sphere (Fig. 1.5).

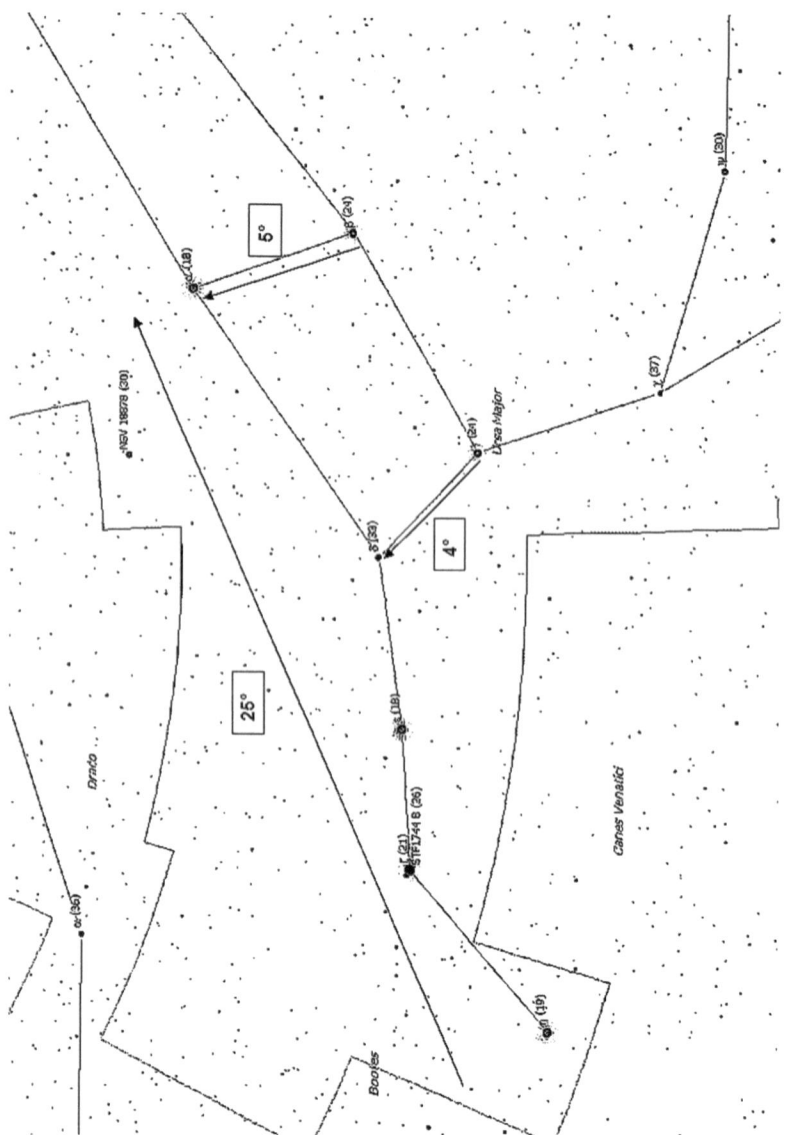

Fig. 1.3 The Big Dipper provides several points of reference that help us to understand distances in terms of degrees. Notice a few common distances are marked in this illustration. (Star chart courtesy of *Sky Tools 3* at www.skyhound.com)

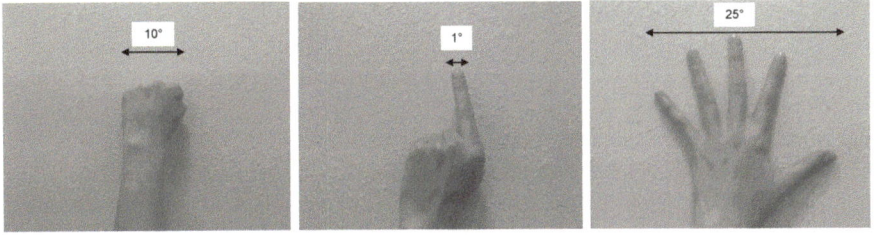

Fig. 1.4 Angular distance can be estimated by using your hand. (Photo by the author)

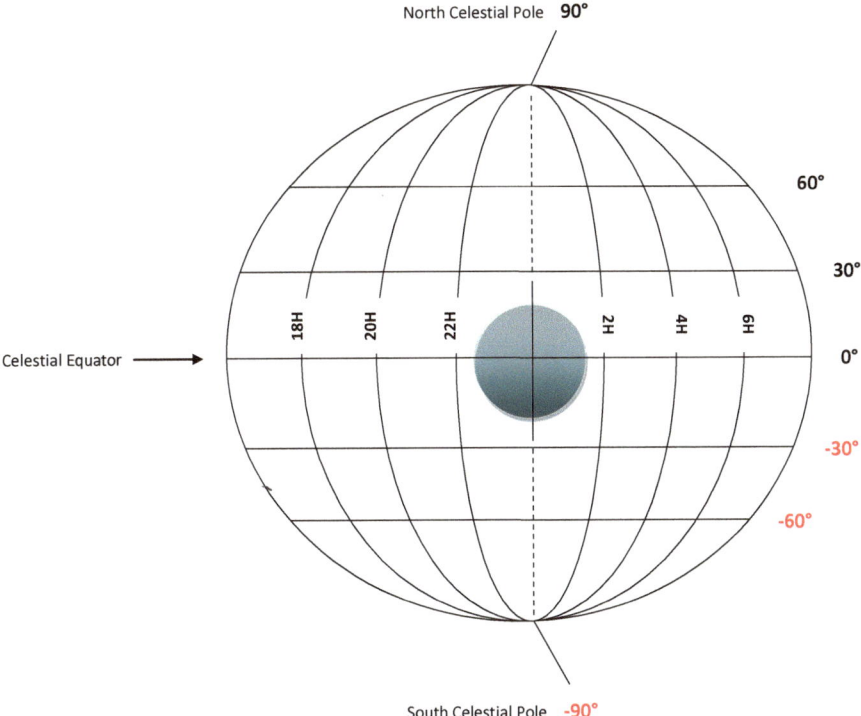

Fig. 1.5 The celestial sphere. Lines of declination run north to south, while hours of right ascension move east to west. Illustration by the author

Mentally project those lines of latitude and longitude outward into space, and now every star has its own coordinates, except that on the celestial sphere we describe these points as intersecting lines of declination (latitude) and right ascension (longitude). Horizontal declination (sometimes represented by the Greek letter δ (Delta) or abbreviated *dec.*) lines descend in degrees from 90° at the north celestial pole to 0° at the celestial equator, then continues to decrease as we move south to −90° at the south celestial pole. Each degree is divided into 60 min of arc, and each minute of arc is divided into 60 arc seconds (s). If you hold up and extend your little finger at arm's length, it is about the distance of 1° across (or 60 arc minutes, written as 60′). Try this and use your finger to cover up the Moon, which you will see is only about ½° (same as 30 arc minutes, written as 30′).

The celestial sphere is also divided into 24 hours (h) that increase when moving in an eastward direction but decrease when moving westward. These vertical lines of right ascension (sometimes represented by the symbol α or abbreviated *RA*) increase from west to east beginning at the point of the vernal equinox, and are divided by hours, minutes, and seconds. It can be remembered by recalling that hours of right ascension *increase* to the *east*, and *set* (decrease) in the *west*. Each line marks 1 h, which is a distance of 15° on the celestial sphere. Every hour contains 60 min of time (60^m), and each minute contains 60 s of time (60^s). To convert minutes of right ascension into degrees just remember that 4 min is equal to a distance of 1°.

Every object has its own coordinates (also referred to as its ephemeris) of right ascension and declination given for a particular epoch or year. If you look up an object's coordinates in an atlas or astronomy software, they are accurate for the epoch in which they are given. Currently we use epoch 2000.0, as any change in an object's position would still be negligible. Significant change occurs over several decades, and so you will usually see published updates in star atlases about every 50 years or sometimes after 25 years. (Precise coordinates for any time can be acquired with astronomy software). These changes come about because of the apparent movement of the celestial objects on the celestial sphere due to precession.

Precession is the continual conical movement of Earth around its axis as it orbits the Sun. It is caused by the combined gravitational effects of the Moon, planets, and the Sun acting upon Earth. As Earth spins on its axis at 23.5°, it completes this circular shape, much like a spinning top, and slowly spins over a period of about 25,800 years. This moves the celestial coordinates on the celestial sphere, thus causing the ephemerides of objects to change. Take as an example the north pole star, which about 4,000 years ago was the star named Thuban (Alpha [α] Draconis), and about 12,000 years from now it will change yet again, when Vega's (Alpha [α] Lyrae) position will be within 6° of the north celestial pole.

How do you go about finding an object described in a text only by its coordinates and the constellation that it lies in? Turn your sky atlas (or look at the example in Fig. 1.7) to the section containing the particular constellation or to the page that contains the range of coordinates in both right ascension and degrees of declination that the object falls within. Carefully look for the hours of right ascension along the top or bottom of the page, and the degrees of declination for the object along each side. Move along each axis until the coordinates on each axis intersect. You should see the object plotted at that location on the map. If it is not there, a possible reason is that it exceeds the lower magnitude limit for objects plotted on that particular atlas. This means that the object you are looking for might be too faint to be plotted on your atlas.

Each sky chart or atlas may have its own method of making it easier for you to locate objects, whether it is the use of running headers or footers on each page containing the range of coordinates covered on each page, or some other kind of key in the front or appendix of the atlas. If you are using a planetarium program on your computer, the object can simply be entered into the database, and it will automatically display the object in its proper location. At that point you can print out that page of the chart containing your destination. Suggestions on some of the available software is discussed in detail later in this chapter under the Section "Constellations." (Fig. 1.6.)

When using a star map you will need to grasp the distances across each chart in proportion to the scale of the map. Each one will be different, as there is no universal size say for example 1° on a star map. However your star map will show along its edges marks showing what distance on the map is equal to either degrees of declination or hours of right ascension. Take a look at Fig. 1.7 of the Sagittarius constellation and you will see along the map's border on top and bottom the hours and minutes of right ascension clearly marked. Also on each of its sides the degrees in declination is shown. Using these references you can quickly estimate that the distance between the α star and NGC 6559 is about 20°. With a more detailed star chart you would be able to see that the distance in reality is 23°. This will give you an idea and perspective of the size of the constellation or distance between points that you are looking for in the night sky. Use the estimated hand measurements of degrees discussed earlier to see how this would look in the sky as you extend your hand.

It is important not to confuse the hours, minutes, and seconds of right ascension (in right ascension there are 60 min in $1\,h$) with the 60 arc minutes contained in a distance of 1°. An angular measurement of 1° contains 60 *arc minutes* (60 arc minutes is written as 60′; each minute contains 60 arc seconds, written as 60″), however 60 arc minutes in degrees does not equal 60 min of right ascension, since right ascension is really an expression of *time*, not distance. To convert minutes of right ascension to be expressed as distance in degrees, always remember that 4 min of right ascension is equal to a distance of 1°, 1 h of right ascension is equal to a distance of 15°, and so on. When using a star chart or atlas use degrees for distance, whether or not you see declination or right ascension scales nearby. However, when specifying an object's location use both its declination expressed by degrees, and its right ascension as expressed in hours, minutes, and seconds.

Direction

Equally important to comprehending distances in either light years or angular measurement, is the correct understanding of celestial direction. The difference between terrestrial and celestial directions will begin to become clear as you see that part of the confusion lies in the difference in directions on a celestial map versus a terrestrial one. On the ground map, north is up, south is down, east is on the right and west is to the left. However, on a star map, while north is up and south is down, west is on the *right* and east is on the *left*. The reason for this is that a star map is printed from the perspective of the observer looking *up* into the sky, whereas a land map is from the perspective of a traveler looking downwards or at least out across the horizon. The placement for each direction is important to remember when consulting any star chart, including the ones provided to help you locate objects described later in this book (Fig. 1.8).

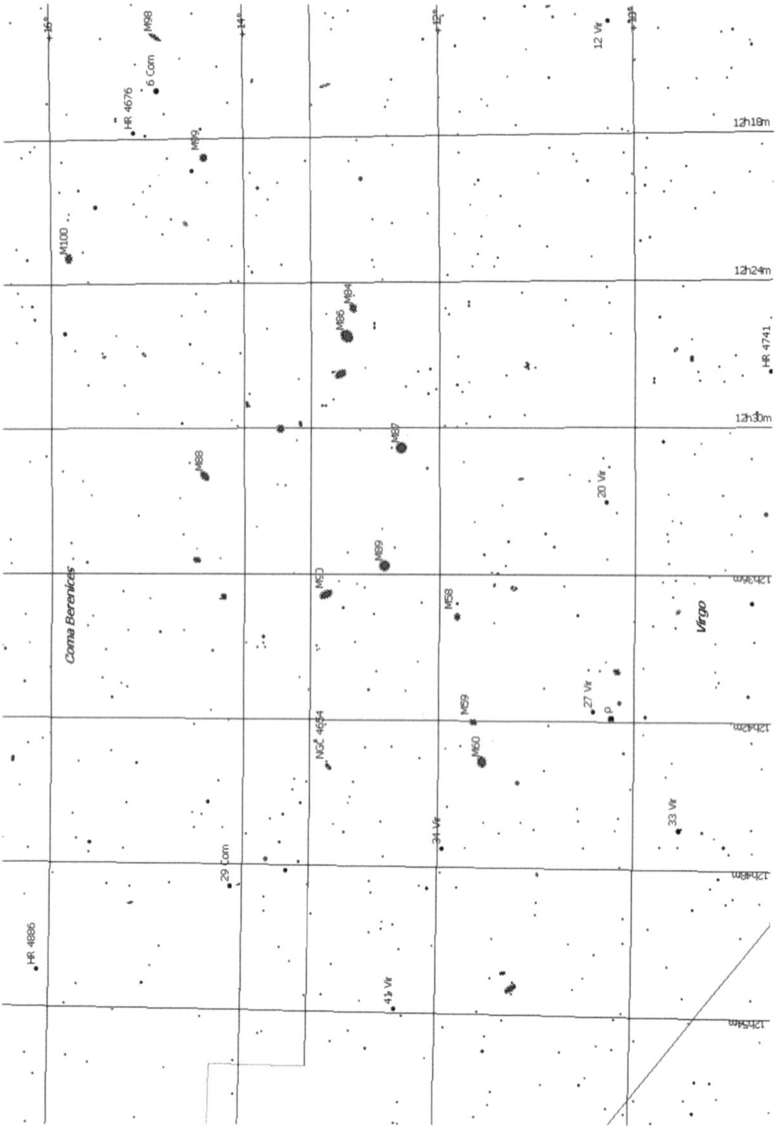

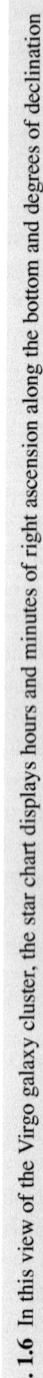

Fig. 1.6 In this view of the Virgo galaxy cluster, the star chart displays hours and minutes of right ascension along the bottom and degrees of declination along the right side. (Star chart courtesy of *Sky Tools 3* at www.skyhound.com)

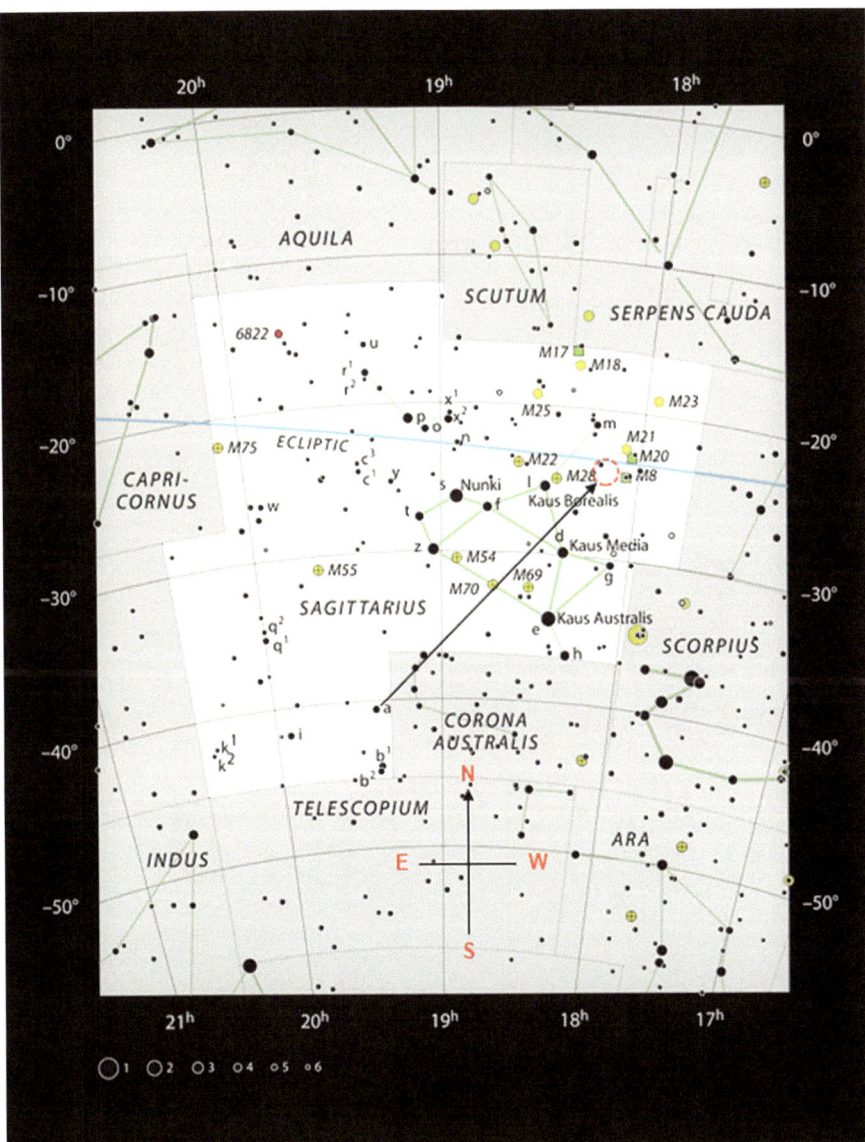

Fig. 1.7 In this map of the constellation Sagittarius, the pointer indicates the location of the emission nebula NGC 6559. (Image courtesy of ESO, IAU, and Sky & Telescope)

A planisphere is basically an all-sky map and a great way to learn the night sky, but you must keep in mind that it is designed to match your view of each horizon. A star atlas or printed map from an astronomy software program will show west to the right and east to the left. One of the best ways to get ahead of this is to simply turn your map to face the direction you are looking in. If you are looking south, a

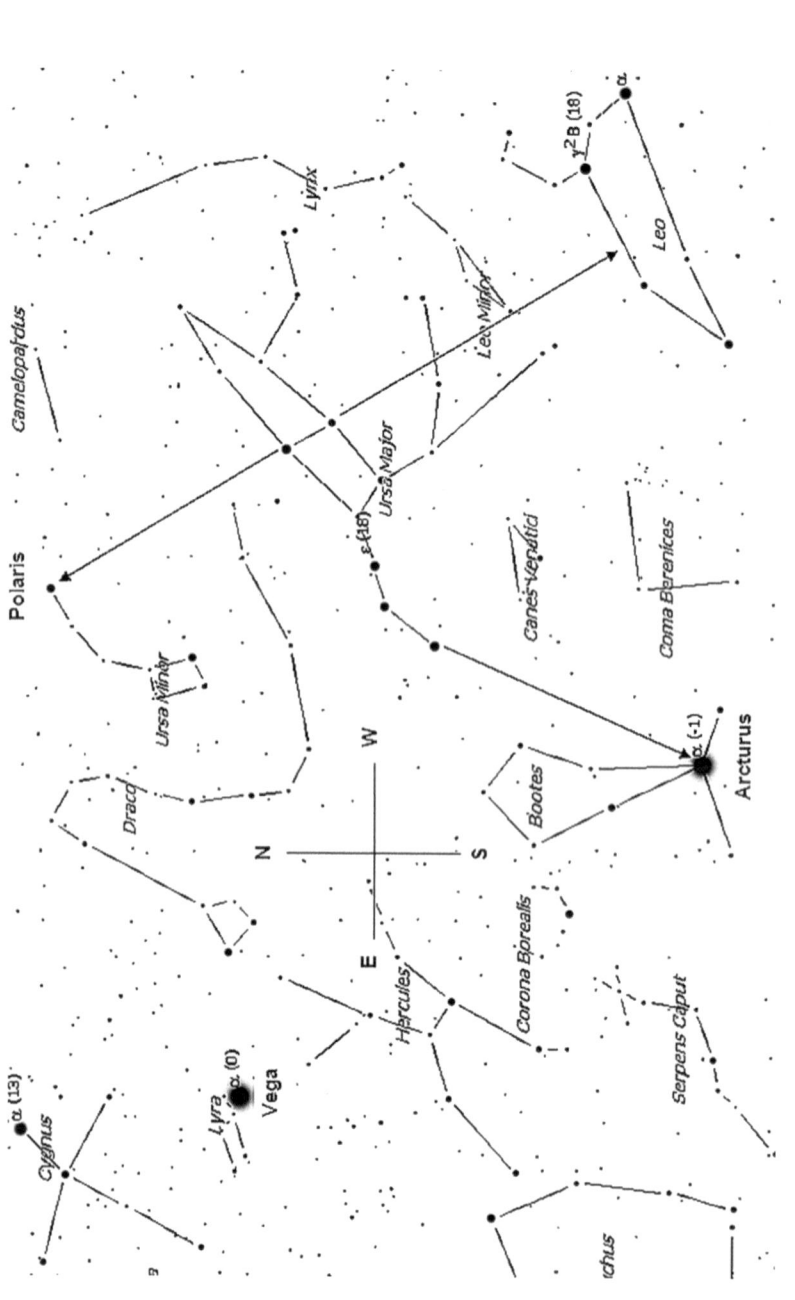

Fig. 1.8 Celestial directions are shown in this map, along with easy to find pointers to bright stars and constellations. (Star chart courtesy of *Sky Tools 3* at www.skyhound.com)

star map will naturally match your point of view, with west on the right side, point-ing in the same direction in which the Sun sets. Raise the map high in front of you. South is down, just as the stars you see in the southern sky, and north is up, stretch-ing through the zenith above your head and backwards to Polaris at the north celes-tial pole. However if you turn around and face north towards Polaris and keep your map oriented the same way as before, it will not make sense. The map will again show west to the right, but since you are facing north, celestial west will be to your left. Therefore, when using a star map or atlas to match your naked eye view, always face the side of the map down into the direction in which you are facing.

Now let's take a few examples to make sure you have this concept well in hand. Take another look at Fig. 1.7, go outside facing south towards the constellation Sagittarius with your book in hand. Sagittarius is an area rich with dozens of beauti-ful celestial gems. If it is summertime this constellation should be visible low in the southern sky. If this constellation is not in the sky, face south and picture it mentally along with the sky chart. Notice the direction indicator; each direction should match exactly with how you are standing in relation to the celestial sphere. South is downwards in the sky, east is to your left, west is to the right, and north extends up and back over your head and behind you towards Polaris.

Now take a close look at the arrow in the illustration that extends from Alpha (α) Sagittae to the object designated as NGC 6559. This is a small emission nebula that requires a telescope to view, so for now if you are looking up in its general area just estimate where it would be using the accompanying map. In which direction does the arrow point? If we are not careful we might say "southwest," but that would be a terrestrial direction and incorrect. After taking a closer look at the chart along with the directional pointers, it is clear that the arrow points to celestial northwest.

In Fig. 1.9 we have two prominent circumpolar constellations displayed in their positions during wintertime, Ursa Minor, which contains Polaris, and also Cassiopeia. Cassiopeia contains many beautiful open star clusters and other objects of interest. On this map, our arrow is pointing from Polaris, which shines at mag-nitude 2.0, towards Beta (α) Cassiopeiae, a 2.3 magnitude star also easily visible to the eye. If you were facing north when using this map, at first glance it would again appear that the directions are mixed up, and it may not make sense. However, when outside trying to match your map's view with the orientation of the stars in the sky it would be practical in this case to rotate the map so that the directional pointers match your view of the sky. The map would be turned upside down, with north pointing directly downwards, and then it begins to make sense. So in which direc-tion is the arrow pointing this time? If you said the answer is *celestial southeast*, you are correct. The arrow is pointing up, behind you towards the south celestial pole, and it is also pointing eastward. By using both the map and sky to visualize direction we can come to the correct conclusion. This also matches the direction indicated by the map itself, which is clear even if it appears we have turned the map around the wrong way. As long as you are guided by the way that the directions point on a star map (north is up and south is down, west is on the *right* and east is on the *left*), along with making sure that the map's orientation matches your current view of the night sky, you will know exactly where you are at.

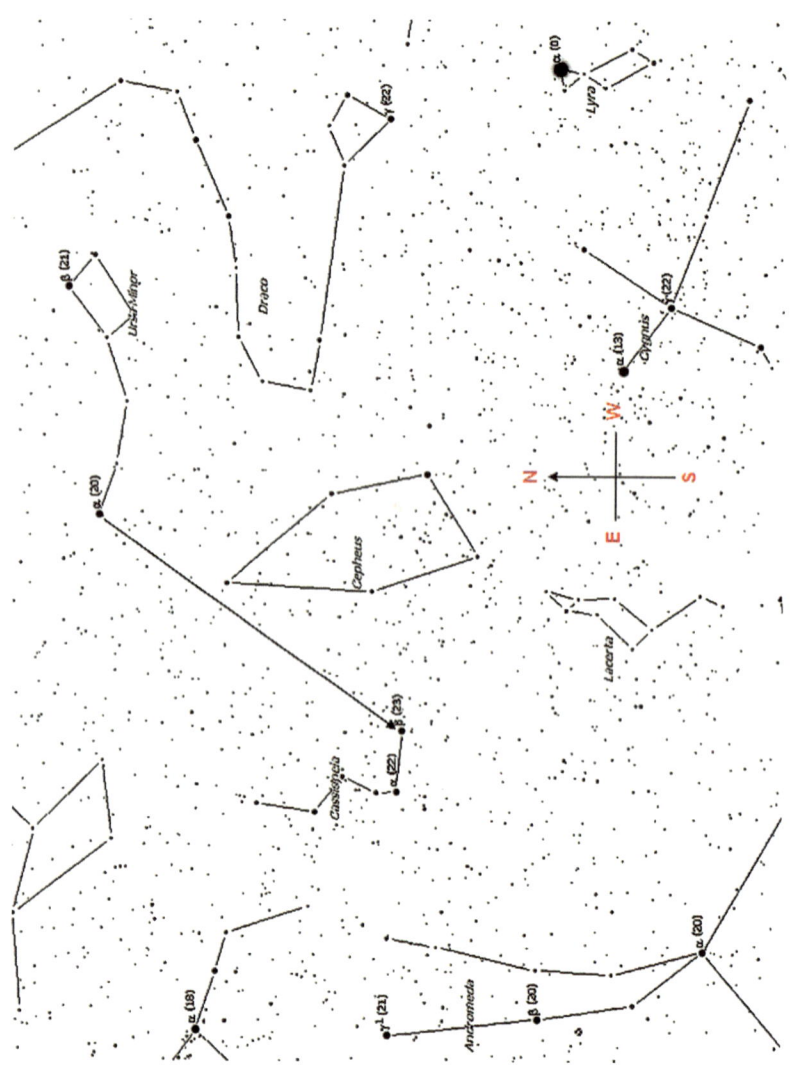

Fig. 1.9 In this example, our arrow points southeast from the star Alpha (α) Ursae Majoris to Beta (β) Cassiopeiae, a constellation rich in open star clusters. (Star chart courtesy of *Sky Tools 3* at www.skyhound.com)

Up until this point we have only been discussing celestial directions as you view them with the naked eye, but it would also apply to the view through binoculars or a red dot finder or other non-magnifying finder. However, directions change when looking through the eyepiece of a telescope. We will cover those details in the upcoming section of this chapter entitled, "How objects will appear in the telescope," but right now let's take a closer look at the effects of Earth's daily rotation and annual orbit around the Sun.

Motion

As Earth rotates, each night we observe that the stars move from the eastern horizon and then set in the western horizon. The rate of apparent movement for the stars and other celestial objects is 15° per hour. That is why you can see some constellations if you stay up late enough to watch them come into view. On the other hand if you enjoy observing objects during January in, let's say, the constellation of Pegasus, if you wait too long this entire group of stars will have set in the west by midnight local time. There are some stars that do not set, but instead rotate around the north celestial pole (marked by Polaris) and are referred to as circumpolar stars. Your latitude marks the distance that Polaris is from the horizon, so if you live at latitude 45° north, Polaris is also 45° above your horizon. If the angular distance from any star to Polaris is less than your latitude, it will not set but appear to circle around the north celestial pole.

Earth also orbits around the Sun, and because of this our view of the constellations changes with each season. This annual movement causes us to perceive a daily shift of the stars at a rate of 1° each day, or, in other words, a star will appear to rise 4 min earlier each night than the previous day. Over the course of weeks this time of 4 min per day accumulates, so we can see why it is that constellations are constantly on the move. Stars that rise at 9:00 p.m. local time tonight will rise 2 h earlier at 7:00 p.m. local time in just 1 month. Possessing this knowledge help us to grasp when are the best times to observe objects, and to anticipate when certain stars will become visible or retreat from our sight.

In addition to apparent motion, a star also has real or proper motion as it speeds along throughout space and moves either towards or away from us. Astronomers determine this proper motion by calculating the angular movement in the sky, adjusting for Earth's rotation around our axis, and its orbit around the Sun. In 1916 famous astronomer E. E. Barnard discovered what is now called Barnard's Star (see Fig. 1.10), an exceptional star because of it having the fastest known proper motion of 10.4″ (arc seconds) per year, or 17.4 million miles (2.8 billion km), which is a movement through our galaxy at a rate of about 89 km per second. It is also interesting to note that this red dwarf star moves at a rate of 20 km per second towards us. This 9.5 magnitude star in Ophiuchus is just 6 light years from us, the next closest star behind the triple star system in Centauri, which is only about 4 light years away.

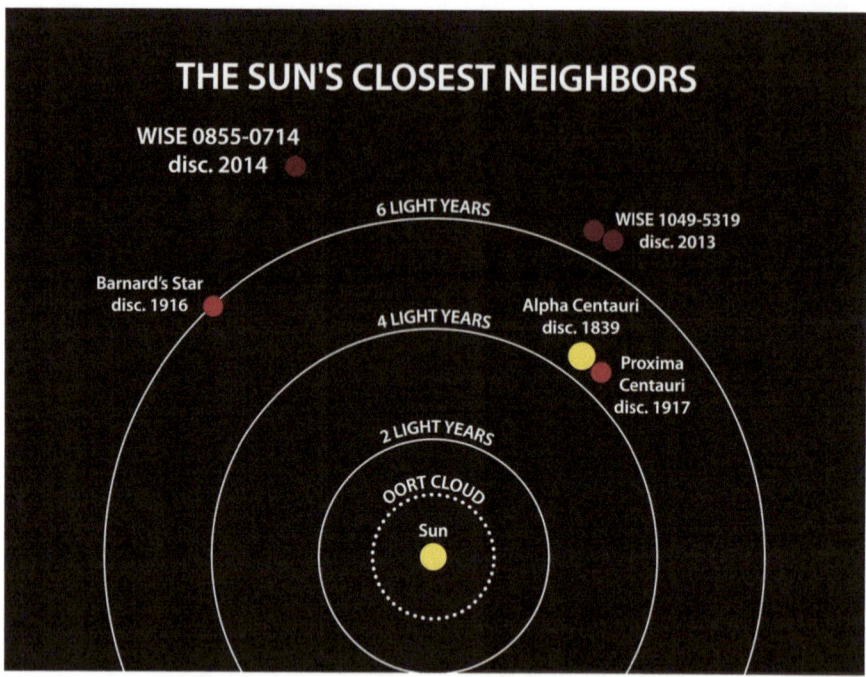

Fig. 1.10 An illustration of the closest stars to our Sun. Listed under each star's name is the year in which our distance from it was discovered. (Image courtesy of NASA/Penn State University)

Star Patterns

Magnitude

In the 2nd century B. C., it was the Greek astronomer Hipparchus who originated the idea that the brightness of each star could be defined by a number. The brightest stars visible to the human eye he called magnitude 1, and next brightest magnitude 2, and down to the faintest star, being magnitude 6. Based on this foundation, astronomers have now standardized references to stellar magnitude so that a star that is 5 magnitudes higher in number than another star is 100 times fainter. Thus, a 2nd magnitude star is 100 times brighter than a 7th magnitude star, and conversely a 7th magnitude star is 100 times fainter than the star of 2nd apparent magnitude. Each magnitude represents a change in the brightness of a star by a factor of 2.512 [$(2.512)^5 = 100$].

Apparent magnitude is different from absolute magnitude. Absolute magnitude (or absolute visual magnitude) refers to the absolute brightness of a star if it were located 10 pc from us. This is simply a standard way of expressing the brightness of a stellar body as compared with our Sun. For persons with extremely keen eyesight, the eyes can perceive stars between magnitude 7.0 and 7.5 in the very best conditions. Of course there will be exceptions in both directions, and some observers have reported seeing objects as faint as 8th magnitude. Using a 50 mm pair of binoculars will reveal stars down to magnitude 10, and a large size amateur telescope of 10 in. or more (254 mm) will reveal stars down to about 14th or 15th magnitude. Even the Hubble telescope has its limit, which is about magnitude 31.

In star charts and sky atlases brighter stars are indicated by a larger illustration of the star, and a proportionately smaller one for fainter stars. In tables listing data for stars, you will usually see the apparent or visual magnitude sometimes abbreviated as "V" or "m_V." Often listed separately is the absolute magnitude (sometimes abbreviated as "M" or "M_V"), which as mentioned above, is the magnitude at which the star would appear if it were located at a distance of 10 pc (32.6 light years). At times there are also references to photographic magnitudes (abbreviated as P or m_{pg}). Photographs tend to be more sensitive to blue light and will therefore result in a star's magnitude appearing differently from a visual observation. Be sure to check the key of the data set you are using to be sure you understand which value is being expressed. In texts like this one, the apparent visual magnitude is usually meant unless otherwise noted.

A practical way to learn apparent magnitude would be to use a monthly star chart, such as found in an astronomy magazine, and compare the magnitudes of the stars listed outside at night as you view each one. This will get you used to discerning the differences in apparent stellar magnitude and allow you to be able to make your own recorded estimates of objects that you observe. Also, at the end of this chapter you will find four seasonal all-sky maps (for latitude 40° North) that will help you begin to navigate your way around the sky in connection with the brighter and more common stars visible during each season. If you are further south you have a better view of constellations such as Scorpius and Sagittarius. Listed below are the apparent visual magnitudes of some of the brighter stars in each season, in ascending order by magnitude:

Fall

Name	Constellation	Magnitude
Capella	Auriga	0.08
Betelgeuse	Orion	0.50
Fomalhaut	Pisces Austrinus	1.16
Alpha (α)	Cassiopeia	2.23
Epsilon (ε)	Cassiopeia	3.37
Gamma (γ)	Triangulum	4.00

Winter

Name	Constellation	Magnitude
Sirius	Canis Major	−1.47
Rigel	Orion	0.12
Aldebaran	Taurus	0.85
Pollux	Gemini	1.15
Castor	Gemini	1.94
Eta (η)	Auriga	3.20

Spring

Name	Constellation	Magnitude
Arcturus	Boötes	−0.04
Procyon	Canis Minor	0.34
Spica	Virgo	1.00
Regulus	Leo	1.35
Mizar	Ursa Major	2.27
Delta (δ)	Ursa Major	3.31

Summer

Name	Constellation	Magnitude
Vega	Lyra	0.03
Antares	Scorpius	0.96
Deneb	Cygnus	1.25
Alpha (α)	Ophiuchus	2.10
Beta (β)	Capricornus	3.08
Kappa (κ)	Lyra	4.33

Stars come in a vast assortment of types with different characteristics such as temperature, age, motion, and chemical composition. Astronomers study these and other properties of stars by using spectroscopy. A star's light can be closely examined by use of a spectrograph. The spectral type of a star is a way of classifying it by temperature and color. You will find many references to stellar spectra in astronomical literature. The hottest stars are a part of spectral class O, while the cooler ones are in class M. The entire order is O, B, A, F, G, K, M. If you need help remembering the sequence you can think of it phonetically as in "Obaf-Gick-Em." Another way of recalling the order is by using an easy phrase where the first letter of each word is one of the spectral classes. One phrase you could use is: **O**rion

Spectral class	Estimated surface temperature
O	30,000 K
B	20,000 K
A	10,000 K
F	7,000 K
G	6,000 K
K	4,000 K
M	3,000 K

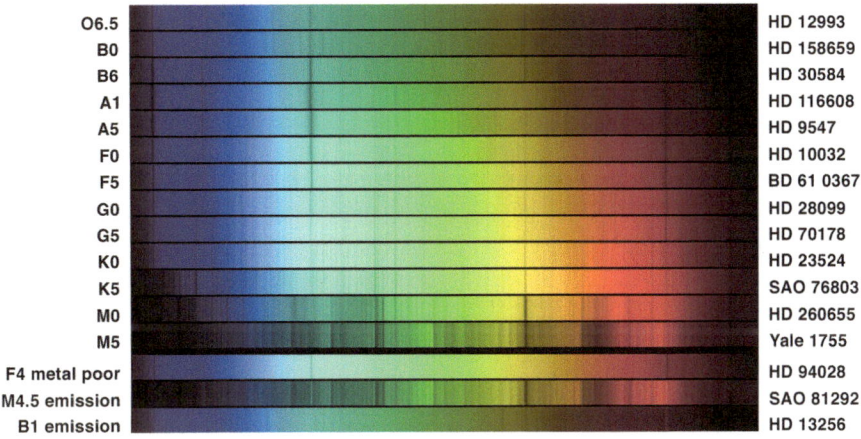

Fig. 1.11 This is an image of various spectra of different stars. Their classifications are shown on the left side and the names of the stars are on the right. (Image courtesy of NOAO/AURA/NSF)

Betelgeuse **A**stronomy **F**or **G**alileo, **K**epler, & **M**essier. Over the years the basic spectral classes have been greatly augmented by other more specific designations, but the basic classes most often referred to can be seen in Fig. 1.11.

Star Designations

Naming of the stars goes back to ancient times, and even today we still have similar names for some of those stars, with the names being translated often from Arabic, Greek, or Latin. Examples are Markab in the constellation Perseus, which comes from Arabic, meaning Saddle, and Denebola in the constellation of Leo, which comes from the Arabic name *Al Dhanab al Asad*, meaning the Lion's Tail. In 1603 a real milestone was achieved when German lawyer and astronomer Johann Bayer created the first star atlas, *Uranometria*, to cover the entire sky. It was in this atlas that Bayer used small Greek letters to label the stars in descending order, from the brightest star being labeled α (alpha) until every letter of the Greek alphabet had been used. Thereafter small Roman letters would pick up where the Greek letters had left off.

In the eighteenth century British Astronomer Royal John Flamsteed (the first one appointed by King Charles II in 1675 for the newly built Royal Observatory in Greenwich, England) created a catalog of almost 3,000 stars, suggesting that they be numbered in order of increasing right ascension from west to east. Although some researchers believe he used this numbering system even while alive, the catalog was published in 1725 after his death. The use of Flamsteed numbers is still in use today.

Stars will often have multiple designations, such as the above mentioned star Denebola, is also known as Beta (β) Leonis or 94 Leonis. When expressing a star's name using either the Flamsteed number or Bayer designation, the Latin genitive (possessive) form is used, so that Beta Leo becomes Beta *Leonis*. As time went on, astronomers began to name stars with numbers based on the catalogs in which they were compiled, several examples are the Henry Draper catalog, the Smithsonian Astrophysical Observatory catalog, and the Tycho-2 catalog. The Tycho-2 catalog of stars was created by the European Space Agency using the Hipparcos satellite, which measured the positions and magnitudes of 2.5 million stars between 1989 and 1993.

In the mid-nineteenth century, German astronomer Freidrich Argelander devised a method of naming variable stars. The first variable star in each constellation is named after a capital Roman letter R, such as R Leonis; the next one would be S Leonis, and all the way through Z, then double letters are used such as RR, RS, RT, and finally RZ. This repeats through the rest of the alphabet until a variable is named ZZ, and then goes back to AA, AB, until arriving at QZ. Any series beginning with J was omitted to avoid confusion with the letter I. Once this series has been used up, astronomers will use a capital Roman letter V combined with numbers.

There is no need to feel intimidated by the various ways in which stars are designated. For use in amateur astronomy, most often people use the common or popular names for stars (such as Betelgeuse in Orion) or the Greek Bayer letter. Begin by learning the names of the brightest stars in the constellations, building on the names of the many stars that you probably already know. Just by looking up the names while stargazing each month, in time you will begin to recall the names without much effort.

Asterisms

An asterism is a group of stars that are shaped or patterned after some recognizable form. For example the Big Dipper for most people is something readily recognized, and this shape is an asterism, not a constellation. The name of the constellation in the northern part of the sky in which we find the Big Dipper is called Ursa Major. Another example is the famous belt of Orion—another asterism whereas the entire constellation spans a much larger area. Sometimes we can mistake an asterism for an entire constellation because it is spread over such a wide area, like the teapot of Sagittarius. However you can see from Fig. 1.7 that the boundaries of this constel-

Fig. 1.12 M45, the well-known Pleiades cluster. (Image by the author)

lation extend much further than the teapot. The next time the Pleiades are up, take a look through binoculars and see if you can detect a prominent asterism in that open star cluster. To many people, it looks like a miniature Big Dipper (Fig. 1.12).

Constellations

In order to learn how to successfully see celestial phenomena through your telescope, it is important to learn the constellations visible in your part of the world and recognize their patterns in the sky. Not knowing the constellation where an object is located can be compared to visiting a city you have never been to. Without preparation it would be difficult to figure out where you are and find your way to important landmarks. You could still enjoy your visit, but your appreciation will be limited, and there will likely be some feelings of frustration. However, if you were told in advance what city you would be going to and you researched places to visit and things to do, your appreciation would be greatly enhanced. A lot of time would be saved since you would have mapped out the best routes to take to your destinations. In a similar way, knowing the constellations will go a long way in helping you to enjoy stargazing.

There are several additional ways to quickly become familiar with the night sky. One is to simply get involved with your local astronomy club. There are usually evenings scheduled that focus on helping beginners learn some of the basics. Allowing someone more experienced to take you on a personal tour of the constellations is effective and also a great way to meet other people who love astronomy.

If you own a smart phone, you can make use of the many astronomy apps available. One such app is *Sky Guide* by Fifth Star Labs. Designed for use with iPad, iPhone, or even iPod Touch, this app gives you access to high-resolution images of the stars visible from your location. Just point your device at the sky and it will identify what you are looking at, and also direct you towards a particular star or object that you wish to find. Tapping an object on the screen will provide detailed information about it and provide access to hundreds of articles. *Sky Guide* can function without Wi-Fi, or GPS. It boasts over 37,000 panoramic photographs to provide you with clear views of how the sky will look and also allows you to control the magnitude of the stars shown on the screen so as to match your local viewing conditions. In addition to displaying the names of the constellations and their stars, this app also illustrates the shape of each constellation's commonly known figures and asterisms. *Sky Guide* is enhanced by its inclusion of great background music specific to what you are seeing, and a red light viewing mode is also featured.

Another great app is *Starmap* by Fredd, available for iPhone, iPad, or iPod touch. It is like having a planetarium on your phone. It uses the GPS technology in the iPhone to identify any area of the sky where it is pointed. The standard version contains 350,000 stars, and 110 deep sky objects. The *Starmap Pro* version is greatly enhanced with 2.5 million stars, and thousands of deep sky objects. It also offers customizable views of the star maps, including the ability to display what your view would look like in both binoculars and in the telescope eyepiece that you specify.

There are also several other excellent options to choose from for an Android smartphone, such as *Sky Map* and *The Night Sky Lite*, both of which are free. The latest full version, called *The Night Sky 2*, has a red light viewing mode and also allows users to send screen shots of their display of stars via email or social media. Simply point your device at an object and the app identifies the object and provides information about what you are viewing. This version is also available on iPhone and iPad (Fig. 1.13).

A much more comprehensive app is *Mobile Observatory*. This is highly recommended if you are an Android user, as it does much more than identify the objects located in your sky. *Mobile Observatory* also provides a calendar of upcoming celestial events, such as an eclipse or a conjunction, and can push them into your phone's calendar to alert you with an alarm. Its database includes 2,500 NGC objects, the entire Messier and Caldwell catalogs (with images), and considerably more. It will also keep you up to date on what is happening with the planets, Moon phases, comets, and minor planets.

Here are a few more tools to help you learn your way around the constellations:

1. You may want to try a constellation field guide. Field guides are compact and easy to carry with you. The National Audubon publishes the *Field Guide to the Night Sky*. It is small and compact, but packs a lot of material within its pages. It is complete with background information on the constellations and is well illustrated. Another classic reference is H. A. Rey's *The Stars, A New Way to See Them*. Although first published in 1954, it was updated as recently as 2008. It is a straight-

Fig. 1.13 Smart phone app Night Sky 2. (Image courtesy of iCandi Apps Limited®)

forward, simple, and clear volume that will help you to learn the constellations with new ways of connecting the stars to form recognizable shapes and patterns.

2. If you enjoy using computer software on your desktop or laptop, there are many good ones available that can help you learn the constellations and plan out your observing sessions. Among those that have received the most favorable reviews are: *Redshift 7, Starry Night, Sky Tools 3, TheSky X,* and *Voyager 4.5.* Each of these features a large object database, high resolution images, interactive tours, and printable star charts. They are also relatively easy to use and include online support with common problems and questions and frequent software updates. There are also planetarium programs available on the Web for free. Although not as detailed as the ones mentioned above, they can still be very useful. One that you may want to explore is *Stellarium* (http://www.stellarium.org). After you provide your location it provides a beautiful, realistic view of the sky in real time

Fig. 1.14 An example of a planisphere. (Photo by the author)

as you would see with the naked eye or binoculars. It will plot deep sky objects on the chart for you, and also allow you to zoom in on portions of the sky and view selected images from its database of the Messier and NGC/IC catalogs. The sky can also be adjusted to simulate the level of light pollution at your location. The program requires minimal computer space and can be downloaded for use with either Linux or Windows.

3. Monthly astronomy magazines (such as *Sky & Telescope*, *Astronomy*, and *Astronomy Now*) include all-sky maps in each issue that are useful for learning the constellations.

4. Planispheres are also very easy to use. A planisphere is essentially an all-sky map that can be adjusted for any time of the year. It is about the size of a standard sheet of paper, and most are very durable. Some can also be ordered specifically for the latitude from which you observe from your favorite online bookstore or may be obtained through local book shops (Fig. 1.14).

After each night that you observe, make brief notes in a notebook about what you saw in each constellation, carefully describing the locations and celestial directions you took to get there. You might also note the date, local weather conditions, and whether or not you observed only with the naked eye or also used binoculars or a telescope. As your notes begin to accumulate, they will show a clear path indicating the progress you have made. Try to spend at least an hour or two outside observing the stars; your eyes will see more stars after the first 45–60 min, and this will let you also get used to seeing the effects of the apparent motion of the stars in the sky. Be sure to look for the more prominent shapes often featured on star charts in each constellation, but do not restrict yourself to only those patterns. Feel free to make up your own patterns. If you see a unique pattern in the sky, it will help you remember it because it will mean more to you than some of the shapes that have been passed down by others through the ages. There are many different ways of connecting the stars into various patterns within the constellations. Some see a bear in Ursa Major, while others see only the Big Dipper, but the important thing is to associate the stars with patterns that you readily recognize within the constellations.

Once you are more familiar with star patterns and constellations, it would also be a good idea to begin using binoculars. There is a certain appeal to their use, as they provide instant satisfaction—you simply point and look. The right side up image orientation you see with your eyes will remain the same in binoculars, except a lot more detail will be revealed. Stars that appeared to stand alone to the eye will show up as double stars. Star clusters will become apparent as round hazy circles in the sky, and even some galaxies will be visible.

Binoculars will also unveil the myriads of stars you see indicated within a constellation on some star atlases but are not visible to the eye because they are fainter than magnitude 7. Clearly, the use of binoculars plays a key role in the development of a keen eye for observing. Once we move to a telescope, however, our field of view gets smaller, and entire constellations are not visible at one time in an eyepiece. So what exactly happens to the appearance of objects when viewed through a telescope?

How Objects Will Appear in the Telescope

From the time of the seventeenth century until now, the quality and size of telescopes has increased dramatically. Even so, the human eye is limited to what it can sense in real time through the telescope eyepiece. Astrophotography uses very sensitive equipment to gather in the light from objects over a long period of time, which in turn creates an incredible amount of detail. More color can be seen in long exposures than with the human eye, but often color in astrophotography is used to create a false color image. This means that the color is simply added to enhance the appearance for either aesthetic purposes or to help make certain aspects of the image more visible by assigning colors to specific features such as clouds, gases, or wavelengths of light that are not visible to the naked eye. In other cases the colors in an image may be genuine, and in such a case, the image is usually denoted as a true color image in the caption. In either case the images are very beautiful, but they do not represent what your eyes will actually see in most amateur telescopes. Even if you could travel to nearby objects in a spacecraft, your eyes would probably not be able to see that level of color saturation since they are not very sensitive to all the wavelengths in the electromagnetic spectrum.

Although long exposure astrophotography is a very important part of astronomy, what you *can* really see with your eyes through a telescope is truly amazing. Visual observing connects us in real time with what we see in the universe. In fact, many amateur astronomers have been able to discover comets, supernovae, and other phenomena while searching the skies with their telescopes. Let's make a detailed examination of just what to expect when looking through the eyepiece of your telescope.

Magnification

As you already know, one important factor in seeing details in astronomical objects is the telescope aperture. The larger the aperture, the more detail you will see, assuming you have a fair amount of observing experience. Magnification is also important as it will also reveal hidden details. However, use high magnification with caution, as this amplifies not only the object but also any poor atmospheric conditions in existence at the time. It is also much more difficult to maintain a steady image under very high power, the image will more easily wobble, potentially decreasing the details you can reasonably observe. To determine the magnification of your eyepiece, note the focal length printed on its side. Then divide the focal length of your telescope by that number. Your telescope's focal length may be printed on it somewhere on its tube or at least will be specified in its manual. Otherwise you can calculate it by multiplying its diameter in mm by its focal ratio (e.g. 152 mm × f/8 = a focal length of 1,216 mm).

Focal length of telescope ÷ Focal length of eyepiece = Magnification.
Example: 750 mm ÷ 25 mm = 30×
Example: 1,500 mm ÷ 16 mm = 94×

Your field of view is also considerably smaller when observing under higher powers which is not necessarily a bad thing, it just depends on your viewing preference or needs for the object. If your objective is to see all of the beautiful wisps of a nebula along with a bright starry background, you will need to use a low power eyepiece because it will provide a wider field of view to accommodate large objects. On the other hand if you wish to see as much detail as you can in the cloud bands of Jupiter, a high magnification is warranted even though it will create a smaller field of view. The general rule of thumb is not to exceed 30–50 power per inch of your telescope's aperture. So if you use a 4-in. (102 mm) telescope, a maximum of 200× is reasonable, if you use a 6-in. (152 mm) telescope—300× is practical. For most needs 300× is practical. For most needs 300× is about as high as you will want to go with the added concerns already mentioned. If you are viewing under the best observing conditions, it is possible to go much higher for objects such as the planets, especially when using a GOTO telescope or a have some other tracking system attached to the telescope. When observing conditions are far less than ideal, you may need to limit your highest magnification to about 25× per inch of aperture.

Field of View and Distance in the Eyepiece

Understanding your eyepiece's field of view will help you determine how much of an area on a star chart you can see. Try using a tool that is easy to make and effective. If your eyepiece gives you a wide 2° true field of view you can mark it down using permanent marker on a small piece of plastic that you would find from the packaging of a household item or a piece can be cut from a sheet protector. You must draw the circle (to scale) so that it corresponds to the same size of a 2° field as shown on your star atlas. Use the degree distance indicator along the sides to determine what size circle you need to draw.

This simple tool will instantly outline the amount of stars shown on your star chart or atlas that will fit into the field of view of a particular eyepiece. Doing the same thing for your binoculars is also a good idea. Binoculars usually provide a field of view of about 5–8°, but check the manual or contact the manufacturer to determine the field of view. What you need to verify is the true field of view, not the apparent field of view (the size it seems to the eye because of the size of the instrument's field stop size). You also can check this by making an observation of the stars of the Big Dipper, and seeing which stars fit within the field of view of your binoculars, just as described earlier when determining estimated angular distance with your hand.

Some eyepiece designers indicate the apparent field of view for their eyepieces, and so with a simple formula you can determine the true field of view. Divide the apparent field of view (as a whole number, e.g., 68° = 68) by the magnification power of your eyepiece and that will be the true field of view in degrees.

Eyepiece apparent field of view ÷ Magnification of eyepiece = True field of view
Example: 50° apparent field ÷ 25 = 2° true field of view
Example: 68° apparent field ÷ 94 = 0.72° true field of view

The challenge in calculating the true field of view is that some manufacturers do not provide the apparent field of view or perhaps it may just be otherwise unknown. You can still calculate it for your eyepiece by timing how long it takes an object to drift across from one edge of the eyepiece field to the other edge. Select a star near the celestial equator, set it at the edge of the eyepiece's field of view, and time how many minutes it takes to drift. Why does it work? The reason is that as we learned before, objects have an apparent motion across the sky of 15° per hour; this would be 1° every 4 min. So then 4 min of drift time = 1° or 60 arc minutes (60′). If it takes a star 3 min to drift across, your field of view is 0.75°; if it takes 2 min your field is ½°, and 1 min is ¼°. This can also be expressed in a couple of basic formulas.

The first is that your field of view in degrees is equal to the number of minutes it takes to drift from one end to the other ÷ 4.

Minutes of drift time ÷ 4 = True field of view in degrees
Example: 4 min ÷ 4 = 1° true field of view (60′)
Example: 2 min ÷ 4 = 0.50° true field of view (½° or 30′)
Example: 1 min ÷ 4 = 0.25° true field of view (¼° or 15′)

The second formula can be used when you prefer to express the drift time in seconds, perhaps because drift time is very fast, such as is the case when observing a planet under high power. In this case we find that the field of view in degrees will be equal to the number of seconds it takes to drift from one end to the other ÷ 240.

Seconds of drift time ÷ 240 = True field of view in degrees
Example: 60 s ÷ **240** = 0.25° true field of view (¼° or 15′)
Example: 30 s ÷ **240** = 0.125° true field of view (⅛° or 7.5′)
Example: 20 s ÷ **240** = 0.08° true field of view (5′)

In the last two examples, the number becomes such a small decimal that it is much more practical to convert from degrees to arc minutes. To do this simply multiply the resulting decimal answer by 60, and you get the field of view in arc minutes as is listed in parentheses (e.g., 0.125° × 60 = 7.5 arc minutes).

These calculations will not only help determine eyepiece field of view but are also good for estimating the angular size of objects that you are viewing. They are also useful for understanding how far one target will appear from another, as described in this book, a magazine article, or in an observing guide. For those that really wish to have more detailed measurements in their observations, there are illuminated eyepieces with double crosshairs that make good points of reference for distance. Going a bit further there are also illuminated astrometric eyepieces that provide several different distance markers in the field of view, making it easier to recall a specific distance that you calculated before. If this interests you, check into obtaining one soon, as they have become more and more difficult to find. For most visual observers, the above mentioned general calculations should suffice.

Direction in the Eyepiece

No matter what kind of telescope you use, if it uses a mirror, the orientation of the image will likely be altered. It will appear different from either the naked-eye view, or perhaps from the way it appears on your star chart. A refractor will flip the image upside down, and if you are using a right angle star diagonal it will also turn the right side of the image to the left, much like a wall mirror does with your image when you look at it. A Newtonian reflector uses two mirrors, so it will read correctly from left to right, but the image will still be upside down.

In a Schmidt-Cassegrain with two mirrors it will function as a Newtonian in regards to orientation, but if you are using a star diagonal, the uneven total number of mirrors will cause it to flip the image from right to left in addition to being upside down. A telescope that has its tube turned so that the eyepiece is facing the right side, may show a different image orientation than one that has its eyepiece facing the left side. From purely a cosmetic point of view this really does not matter to most people; the stars will still be just as beautiful no matter which way the image is turned. However it is good to be aware of the change, as the view of objects in your telescope may differ from the orientation you see in images printed in publications, as well as different from the view in a friend's telescope.

Aside from image orientation, knowing your celestial directions while observing through the eyepiece is a vital part of making it enjoyable and will prevent confusion or uncertainty. Just as it is important to understand directions on the celestial sphere and how that translates to what you see in the sky or on a star chart, it is necessary to understand how direction is affected in the eyepiece. Moving from one target to another can become confusing if you are unsure whether to move up or down when your map indicates south. Or if an article in your favorite astronomy magazine says to begin by centering the star Regulus in your field of view, and then move about 9° east to find the spiral galaxy M95, would that be left or right in the eyepiece? Let's make this easy. Here is the way to get the directions straight no matter what orientation you have in the eyepiece.

If your telescope provides a correct left to right image, such as the case with a reflector with two mirrors or a refractor without a star diagonal, then use the following steps to determine celestial directions in your eyepiece:

1. First center your target and find east and west. Objects will *enter* your field of view from the east and *withdraw* to the west.
2. North is always in the next position *clockwise* from east.
3. To confirm, be sure that the word N E W is spelled in a *counterclockwise* direction.
4. The missing point would of course be south.

If your telescope provides a reversed image from right to left, as in the case of a refractor or a Schmidt-Cassegrain with a star diagonal attached, then use the following steps to determine celestial directions in your eyepiece:

1. First center your target and find east and west. Objects will *enter* your field of view from the east and *withdraw* to the west.
2. North is always in the next position *counterclockwise* from east.
3. To confirm, be sure that the word N E W is spelled in a *clockwise* direction.
4. The missing point would of course be south (Fig. 1.15a, b).

So when you are comparing a star chart to the sky as you look with your naked eye or through binoculars, simply orient the chart towards the celestial direction you are facing, as described earlier. When the moment comes to view the image through the eyepiece, if you are using a telescope that provides a correct left to right image, just rotate the map around 180° to see how your view would be oriented through the eyepiece, perhaps also using the suggested field of view indicator made from a plastic square with your eyepiece field of view marked on it. If your telescope gives a reversed image from right to left it will not be possible to turn the chart for an exact match to the eyepiece orientation, but using the steps given above, you will know the directions in your eyepiece and be able to navigate successfully.

For some amateur astronomers, having a star chart that matches the view in their reversed image telescope is a must. Most star chart software programs will give you the option of printing a star chart with a reverse image view, to match a telescope that reverses the image from right to left. Keep in mind that the actual view in your 'scope may be a little different, depending on the orientation of the tube itself, and which side of the 'scope your eyepiece is situated on. The key is to turn your map around until it matches the view in your eyepiece.

a

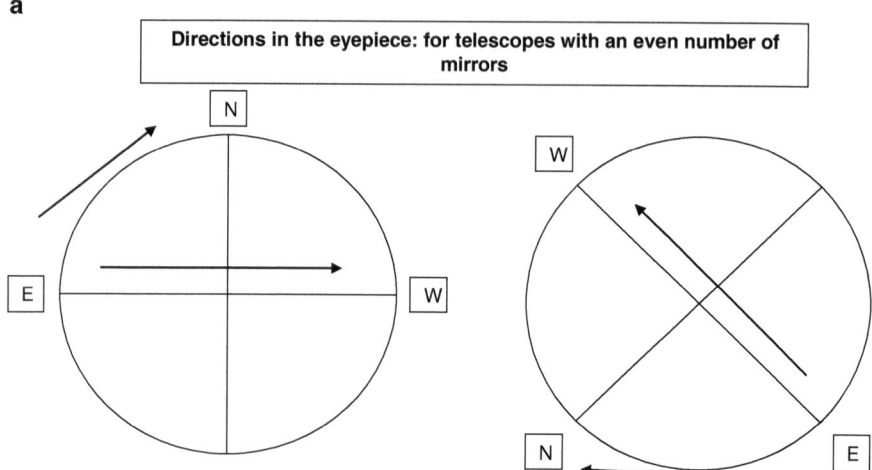

1) Observe the direction of drift so you can determine east and west.

2) Remember that north is always in the next position clockwise from east. The word NEW is spelled in a counterclockwise direction.

3) The missing point is of course south.

b

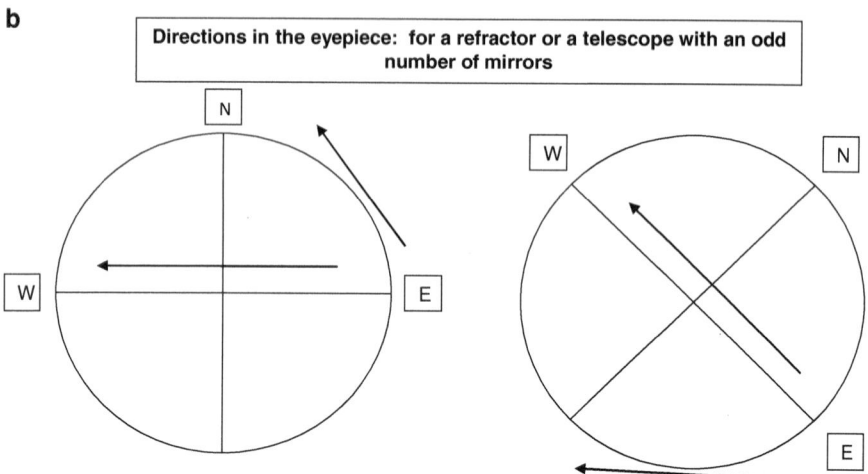

1) Observe the direction of drift so you can determine east and west.

2) Remember that north is always in the next position counter clockwise from east. The word NEW is spelled in a clockwise direction.

3) The missing point is of course south.

Fig. 1.15 (**a, b**) Directions in the eyepiece. (Illustration by the author)

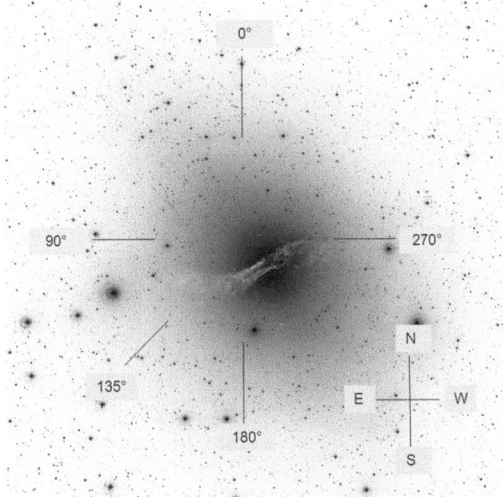

Fig. 1.16 In this image note the various position angles that are referenced. Notations have been added by the author. (Image courtesy of ESO)

Position Angles

At times we need to specify at what point a certain feature appears within a 360° circle (such as in an eyepiece field of view) or we need to describe in which direction an object is pointing from one end to another. This is more precise than simply stating for example that a galaxy appears elongated from northwest to southeast, or that there is a knot on the southeast spiral arm of the galaxy. Each circle has 360° and position angles begin at 0° for north, and increase in a circular direction first to the east, which is 90°, south is 180°, and finally west is located at 270° (see Fig. 1.16).

In addition to this method of description, sometimes the words "preceding" and "following" are used as in the case of some of the descriptive notes in the NGC catalog. The term preceding refers to west, and following refers to east. As an object such as a galaxy drifts across your eyepiece, it is moving west and so an object that lies ahead of it would be referred to as preceding it, moving west ahead of it. On the other hand a star or an object that lags behind, crossing the field of view *after* the galaxy is already moving west out of your field of view, would be following it, entering from the east. So the term preceding is sometimes used to for west, and following for east (see Fig. 1.17). The images therein have been enlarged for easier viewing. The galaxy in both images is Centaurus A, which can be detected in binoculars. Good views can be obtained in amateur telescopes of at least 4 in. (102 mm), and stunning views are possible in larger 'scopes of at least 10 in. (254 mm). You will have to be at least as far south as latitude 30° North to be able to see it, although it will not appear as pristine because of haze near the horizon.

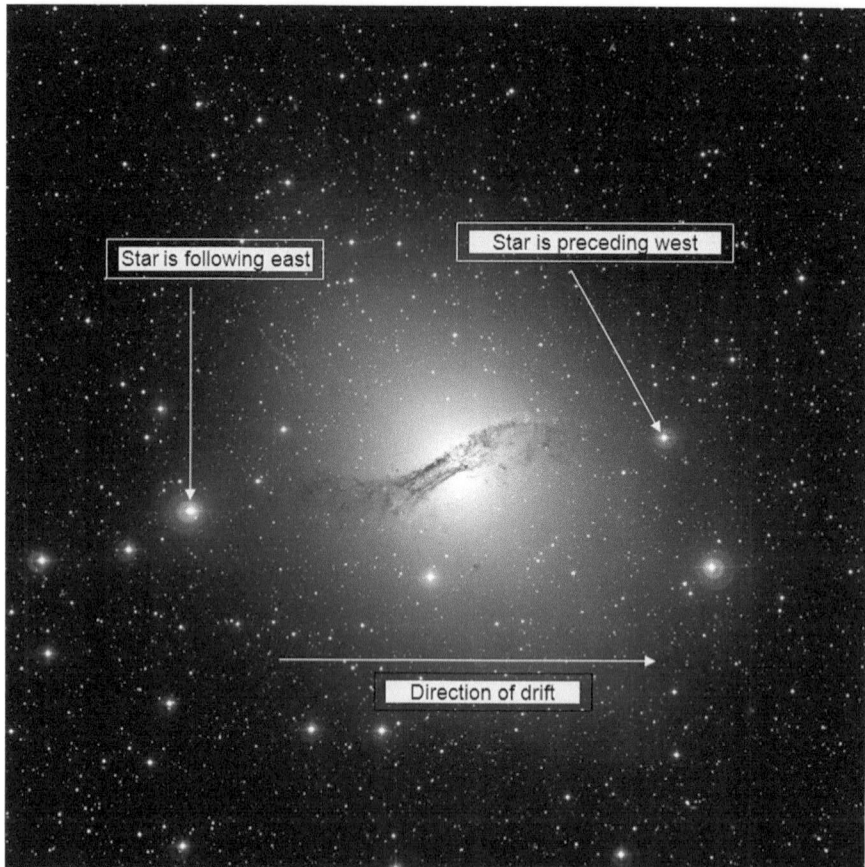

Fig. 1.17 Notations in this diagram were added by the author. (Image courtesy of ESO)

What You Will See

If you have not already spent a lot of time viewing celestial objects through a telescope, prepare to be amazed! Imagine that little cloudy area that you just barely detect with your eyes coming alive in the center of your eyepiece, with hundreds of gleaming stars streaming away from its center. Or seeing a hazy patch, seemingly smaller than the last, but upon visual inspection you will realize that it is an enormous star city, and in fact the nearest galaxy to ours. Not only do you see its large disc, but also its two companion galaxies beside it.

Turning to the winter skies to the constellation Orion, you will no doubt be very excited to see the familiar wispy shape of the Orion Nebula. Some observers are able to see some color in it, often describing it as greenish in appearance. More than that, a modest telescope should be able to resolve (distinguish) the four stars called the Trapezium, located within the Orion Nebula.

These are just a few of the celestial treasures that await you. So much can be observed even with a 'scope that is just 6 in. (152 mm) in diameter. Never underestimate the capability of your telescope to show you an object or feature that is supposed to be out of reach for a telescope of your aperture. Many times it has proved true that with experience, multiple observations, and patience during observing sessions, a difficult object or feature can finally be located and perceived. Sometimes it is a matter of waiting for a night that has better seeing conditions. With careful preparation and experience, your telescope will be able to open up views within our universe that you will never forget. Several of these celestial phenomena will be revealed in Chaps. 4, 5, and 6. For now, here are a few more important details that you should be aware of in order to have successful observing sessions.

Proper Setup of Your Telescope

The proper setup of your telescope will help eliminate disappointment during your observing experiences. Collimation is often an overlooked requirement to a successful night out with your instrument. There is also the question of what to do about dust on your optics. If you wish to enjoy the benefits of your equatorial mount, polar alignment is necessary. Also, as you may already know, you will locate targets much more efficiently if your telescope is aligned with your finderscope. Let's consider these topics one at a time, beginning with collimation.

Collimation

Collimation is a step that you should not skip. Your 'scope's manual should have some instructions on how to collimate, and here are a few additional guidelines. If you are using a refractor it has already been collimated at the factory and may not need any adjustments for years to come. In fact, many refractors are not designed for the user to collimate, and so must be sent back to the manufacturer for adjustment.

For those who own Schmidt-Cassegrain telescopes, collimation is adjusted with a set of three screws on the back of the secondary mirror at the front end of the scope. Night air will be quite cool, so make sure that your telescope has been outside for a while and has had enough time to reach the same temperature as the outside air. Begin by centering a star in a medium-powered eyepiece. Defocus the star, and you will see a dark spot (the shadow of the secondary) within the bright defocused star image. Make slight adjustments to one screw at a time until the dark center dot is right in the center. Be aware that during this adjustment process the image of the star may no longer be centered, and so both the defocused star and the dark center dot will have to be adjusted repeatedly. Next, perform the same process under high magnification. While out of focus the star should show a set of rings surrounding it. Keep adjusting the collimation screws until the star and its rings all line up in a uniform fashion. As a final check, bring the star into focus. Any diffraction rings around the star should also be concentric.

The process for collimating a Newtonian reflector is similar. Some reflectors come with what is called a collimation cap. This looks like a dust cap for the focuser with a tiny hole in its center. If your reflector does not have one you could simply poke a tiny hole through the dust cap with a similar effect. Begin by making sure that the secondary mirror is centered underneath the focuser and the mirror itself should be rotated so that it is evenly pointed directly upwards through the focuser. If it is not, carefully adjust the central screw on the rear of the secondary mirror to adjust its position until it is directly centered underneath the focuser.

The next step is to adjust the vertical tilt of the secondary mirror until you can clearly see the entire reflection of the primary mirror. Use a small Allen wrench to adjust the set of three (sometimes four) collimation screws on the back of the secondary until is it properly aligned. This reflection should show the circle of the primary, centered within your view of the secondary mirror as you look through the focuser (and through the collimation cap if you are using one). The last step is to adjust the collimation knobs or screws located on the back of the primary mirror. Keep turning one screw (or knob) at a time until the image of the secondary mirror is right in the middle of the entire view as you look back through the focuser. If you are using a collimation cap, its tiny dot should be right in the center of your view and everything all together should resemble a bull's-eye (Fig. 1.18).

Although opinions on the use of laser collimators vary, they do work well for collimating Newtonian reflectors. The cost is both reasonable and well worth both the precision and savings in time. When you adjust the secondary mirror to be right under the focuser, the laser helps speed up this process because once you turn it on you will see a red dot right in the center of the secondary. If you do not, then the secondary is not properly centered under the focuser. To help with this part of the task, you could center mark your secondary mirror with a very tiny mark, but with a sharp eye it should be fairly easy to tell if it's centered.

After you are certain that the red dot of the laser falls right in the center of the secondary, look down the tube towards the primary. Another red dot should be projected onto the primary mirror. Adjust the collimation screws at the back of the secondary mirror until the red laser points directly into the center mark on the primary mirror (usually either a black dot or a small hole that looks like a paper hole reinforcement). The last step is to adjust the rear mirror cell screws. With the use of a laser, you do not need to go back and forth from the back of the scope to the focuser, because many lasers are now designed with a viewing window tilted at a 45° angle. Simply adjust the primary mirror collimation screws until you see the red laser beam (which will appear as a fuzzy star) disappear into the center black hole of the viewing window.

A few points of caution: (1) Be sure that if you use a laser collimator, tighten it inside the focuser just as you would an eyepiece. (2) Collimating involves the use of tools, so be careful not to drop anything onto any of your mirrors. Keep the tube

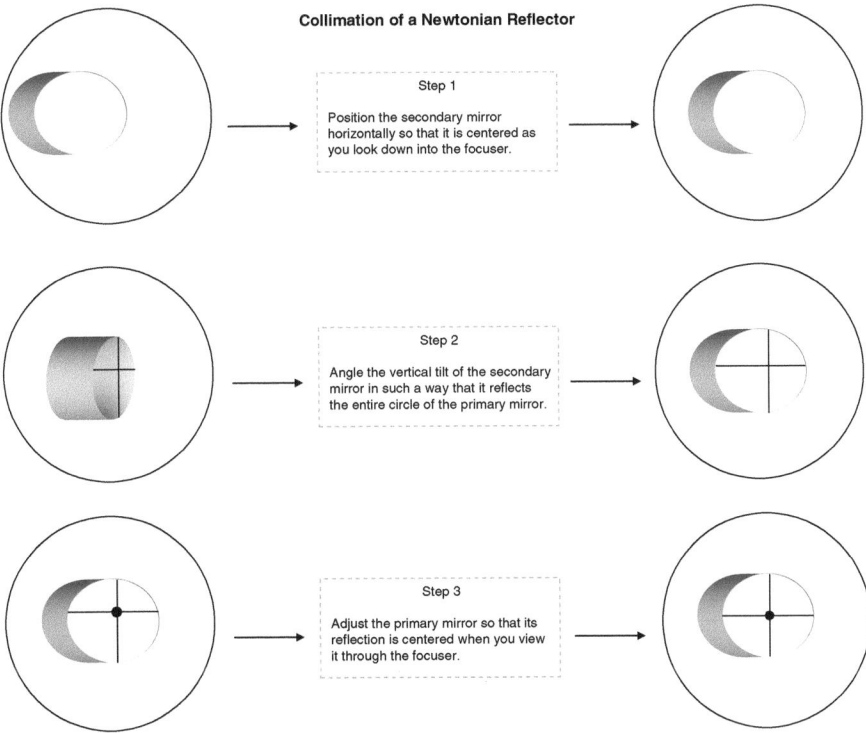

Fig. 1.18 Collimation of a Newtonian reflector. (Illustration by the author)

in a horizontal position to limit the possibility of something falling onto the primary mirror. (3) The poor images seen by observers using their telescopes for the first time, can often be traced back to insufficient collimation. Take a few minutes to practice this essential skill and you will greatly enhance your view of the night sky.

Cleaning the Telescope

The best way to keep your optics clean is to prevent dust and contaminants from reaching them in the first place. Despite the best of intentions, after a long time of use, you may find that your telescope's optics have become a little dirty. There are many, many differing opinions about cleaning optics, from the type of cleaners and tools to be used to how often the optics should be cleaned. However, you would probably agree that the old sayings, "If it is not broken do not fix it," and "Better safe than sorry," both apply in this case.

First and foremost, store your telescope in a cool dry place, away from areas that tend to collect dust and dirt. Second, keep your equipment stored in the containers they came in, and you will usually find that dirt on your optics will be minimized. Third, keep the outside surfaces clean by wiping with a clean damp cloth. Removing dust from the outside will minimize dust from getting on the optics inside. Next, always be careful not to touch the surfaces of lenses or mirrors with your fingers. If your telescope is new, it is unlikely that you will need to clean it for many years.

The wise course of action would be to not rush into a cleaning. Some telescope owners become anxious over a moderate amount of dust, and although no level of dust is welcome, the truth is that a little dust will not ruin your image. For the needs of an amateur astronomer, it would take a considerable amount of contaminate to degrade an image significantly.

On the other hand, if you purchased a used telescope it is quite possible that it was never cleaned and may have accumulated a lot of dirt after many years. So if you believe the optics absolutely must be cleaned, be gentle, and never use harsh strokes or apply strong pressure. When cleaning lenses, the use of a bulb air blower and a camel hair brush to remove dirt is usually sufficient. In some cases you may have to wipe the surfaces gently with a microfiber cloth.

For mirrors, first use a bulb air blower *only*. In most cases it can remove those stray bits of dirt that seem to inevitably land on a mirror over time. If it needs a little more thorough cleaning, you may need to use water. Rinsing with tap water leaves water spots and can leave behind minerals that are just as bad as the dust you are trying to remove. Instead, gather several gallons of distilled water and gently pour them over the mirror, and that should be enough to remove the more stubborn bits of dust. Distilled water will not leave any spots on your mirror. In an extreme case, a *little* drop of dishwashing liquid can be added to one gallon of distilled water, then pour onto the mirror and let it sit for a few minutes. Rinse as described above, and let air dry in a room that is kept very clean. The mirror is best left to dry tilted on its side, perhaps leaning up against a wall or inside a wide, sturdy plastic container. Be careful that it does not slide around. Placing towels underneath and behind it may help to prevent this.

Polar Alignment

A telescope that is on an equatorial mount is capable of allowing you to more easily keep an object in the field of view despite its apparent motion. When the telescope is polar aligned (so that its axis is parallel with the north celestial axis), for the most part you will only need to make an adjustment to the right ascension in order to follow an object as it moves west through the sky. Before beginning the physical process, take a few minutes to look up the coordinates of several bright stars that will be visible in the night sky. You can find them in a star atlas or using an astronomy software program. Choose several that are not near Polaris. Once you have done that, go outside and make sure that the telescope is on a level surface, and adjust the legs of the mount until the 'scope is level. You can use a bubble level tool if your 'scope does not have a level built into its mount.

Next unlock the latitude lock and adjust the latitude setting to match your latitude (e.g., 40° North). Once you do this, you will not need to change it again unless you use this 'scope to observe from a different latitude. If you are unsure of your exact latitude, it can be found in an atlas or on the Internet. Target the star Polaris in a low power eyepiece. Then, unlock the declination dial and set it to 90°. Also unlock the right ascension dial and set it to 02 h, 50 min. Carefully watch the star over several minutes. Polaris should remain centered in the field of view. Now move your 'scope to a familiar bright star that you noted earlier, at least 30° away from Polaris. Adjust the setting circles for both declination and right ascension to match the known coordinates of the star. Now turn your 'scope back to Polaris, and the dials should reflect the known coordinates for Polaris. They may be slightly out of adjustment, so move them to indicate the correct coordinates and then go back to the other star and verify that your 'scope's setting circles show its correct coordinates. All of the objects have positions in relation to one another on the celestial sphere. Repeat this process with one more bright star from your list.

Now it's time to test it on an easy target. Try an object that is in a constellation that is familiar to you, such as Triangulum or Andromeda if you are observing in the late fall or early winter. The coordinates for the galaxy M33 are right ascension: 01 h 33 m 51.3 s, and declination: +30° 39' 54". Move your telescope first to its declination, and lock the axis in place. Then slowly turn its right ascension and you should be fairly close to seeing it within your field of view. Since you are using an object in a constellation that you recognize, you will be able to easily see if you are pointed in the right direction. Also, choosing an object such as M33 that is visible to the naked eye as well as within reach of binoculars will help confirm the identification of what you see in the eyepiece. Once your 'scope is polar aligned, you will be able to center objects in your eyepiece after retrieving a target's coordinates from an astronomy app or a star atlas. Then move each axis of the telescope to the correct coordinates.

Alignment of the Finderscope with the Telescope

The finderscope is an indispensable tool that you will probably use each night you observe. Much like having a pair of binoculars attached to your telescope, it offers a wide field of view and at the same time provides just enough additional light-gathering power to see the detailed patterns that fainter stars create. With it you will be able to target specific objects, and then be able to find those same objects in the eyepiece of your telescope immediately. Without it, it would be extremely difficult to aim at a point in the sky and capture such a small target (often just a few arc minutes across) in a 1° field or less within the eyepiece.

Finderscopes come in a variety of forms, such as a regular magnifying finder-scope, or a non-magnifying reflex sight or illuminated red dot finder. Many observers love to use the famous Telrad, which is essentially a very effective illuminated bull's-eye (see Chap. 3 for more details). Whatever your preference may be, you must be sure to take the time to align it with the telescope.

The process is fairly simple and does not take a lot of time. During the daytime (of course NEVER aim towards the Sun or even anywhere in its direction), find a faraway target such at the top of a telephone pole, flagpole, treetop, or a similar target, and center it in a low-power eyepiece. Next you must go to the finderscope and slowly turn the adjustment screws one by one. On magnifying finderscopes there can be anywhere from three to six screws. On red dot finder there are usually two or three adjustment knobs. There is usually some kind of centering aid in the finderscope such as crosshairs or a red dot.

As you turn each adjustment screw, note how each one causes a change to the location of the target that you chose. Keep adjusting the screws until your target is right in the center of the finderscope. Now go back to the eyepiece of the telescope and make sure the object is still centered (you may have inadvertently bumped the 'scope during this process and shifted the target off center). Adjust it if necessary and then readjust the finderscope. Once night falls, repeat this process with the first star that becomes visible, again using a low-power eyepiece. The more precise you are with this alignment the faster you will be able to locate objects that you seek with the finderscope.

Summary

Here are some questions to focus on:

- Which essential concepts will lead me to being a successful observer?
- How will Earth's rotation affect the apparent motion of the stars tonight? How may it affect my observations during the time I have selected for my observing session?
- What is the celestial sphere, and how can I learn my way around within it?
- Why is it a good idea to develop a good understanding of distances, including angular measurement?

Becoming a successful observer involves developing the habit of exploring the night sky. Perform naked-eye observations as often as you can, and record what you see. Which constellations do you recognize readily and which ones are still unfamiliar? Commit to filling in the gaps.

On the celestial sphere, distance is expressed by angular measurement. One degree is divided into 60 arc minutes, and each arc minute contains 60 arc seconds. Refer to Fig. 1.4 for some common examples. When reading descriptions of celestial targets that involve angular measurement, you may wish to enhance your understanding by referring to your own star chart. Examine the declination scale on your star chart, and note what the distances are between various points that are referenced in those descriptions.

Celestial direction is different from the way we think of directions on land. North is always in the direction of Polaris. Thinking in terms of a sphere should help keep it clear in your mind. On a star map north is up and west is always on the right, east to the left, and south is down.

Knowing the names of the more prominent stars will help you to become comfortable with navigating through the sky. As important as this may be, there is no need to attempt to memorize a long list of stars. Instead, simply note the names as you read about them, and try to refer to a few stars by name each night you observe. Over time the names will begin to stick with you and eventually become second nature.

Here are a few additional reminders and points to keep in mind:

- There are a wealth of resources available to help you become familiar with the constellations that are visible in each season of the year. Whatever format you prefer, whether print or digital, will help *you* to become the expert instead of leaving your tour of the celestial sphere solely in the hands of a computer.
- Be sure that you understand the magnification of your eyepieces and their true field of view. Some observers like to put a label on the barrel of the eyepiece with that information. Another suggestion is to list the information on an index card or a sheet of paper, and keep it with your sky charts or atlas. Laminating it would be a good idea so as to protect it from dew commonly found in the night air.
- Clean your telescope's optics only when necessary. Never use harsh cleaning methods or products that could scratch or damage the optics.
- Collimation is a vital part of ensuring that your telescope will transmit the best possible images into the eyepiece. It requires patience but only involves a few steps to complete. If you feel that you need another set of eyes to make sure your 'scope is properly collimated, a star party or astronomy club meeting is a good place to find someone to give you further assistance.

Enjoying first light with your telescope is an exciting time, but even more thrilling is going beyond first light and learning to enjoy all that visual observing has to offer. So far we have reviewed several of the basics, and by mastering them you are sure to become a successful observer. In the near future this may lead to your deciding to specialize in a certain area of amateur astronomy such as planetary observing, astrophotography, or searching for supernovae. In view of the prevalence of light pollution, in the next chapter we will take a look at where you can find the best viewing of the night sky. We will also discuss when are the best times to view specific types of objects. In the meantime, the stars await you, and amateur astronomy is all about *your* exploration of the universe. So grab your 'scope, get outside, start observing, and rediscover why you fell in love with the universe (Fig. 1.19a–d).

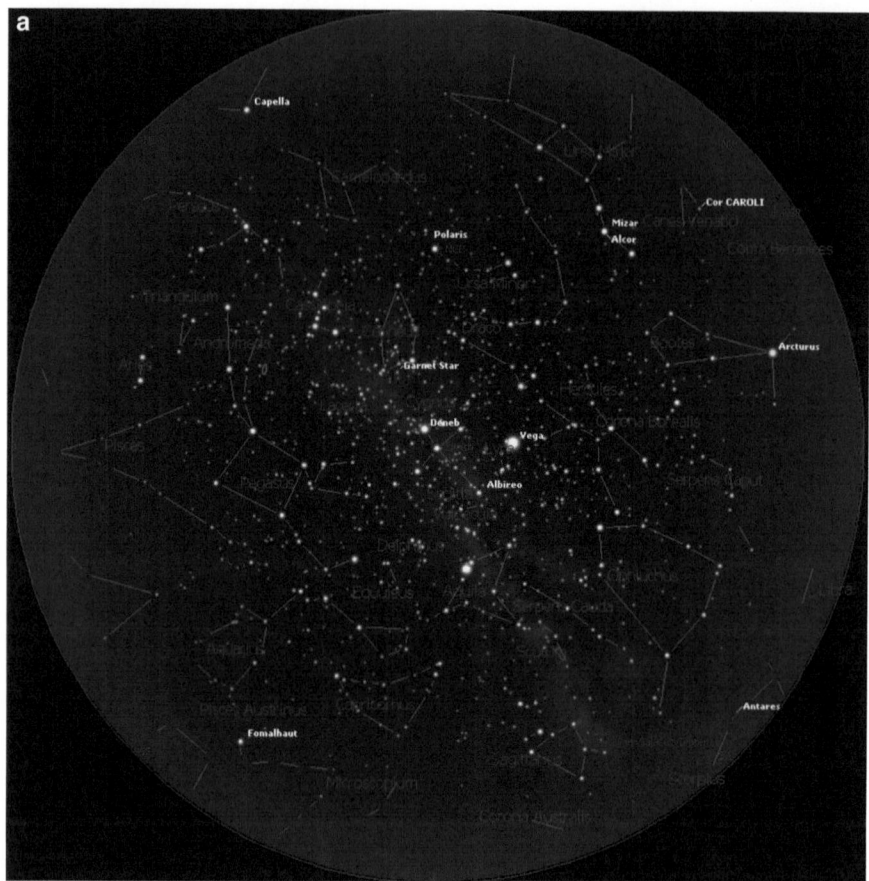

Fig. 1.19 (a) Fall. (Star chart courtesy of *Sky Tools 3* at www.skyhound.com.)

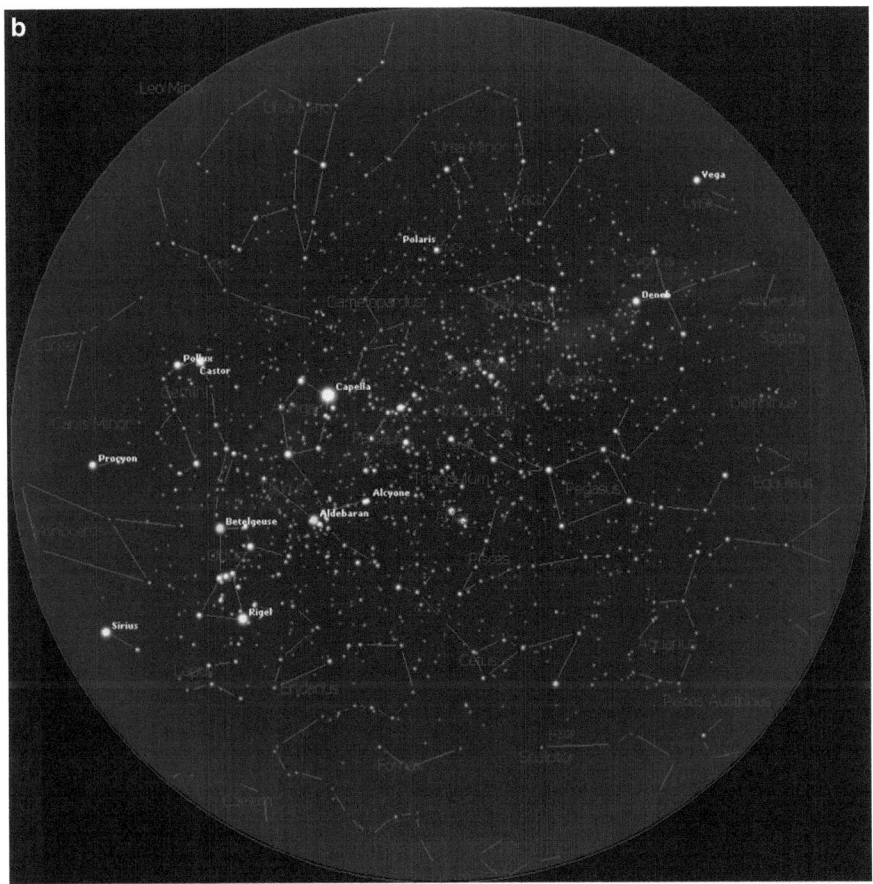

Fig. 1.19 (continued) (**b**) Winter. (Star chart courtesy of *Sky Tools 3* at www.skyhound.com.)

Fig. 1.19 (continued) (**c**) Spring. (Star chart courtesy of *Sky Tools 3* at www.skyhound.com.)

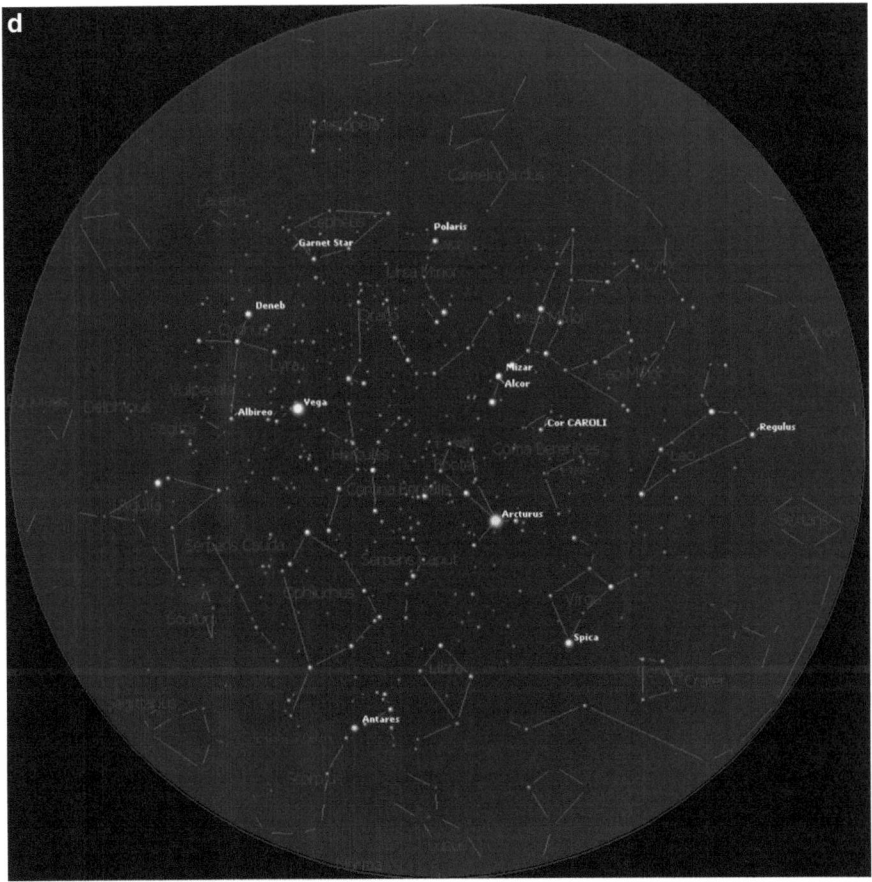

Fig. 1.19 (continued) (**d**) Summer. (Star chart courtesy of *Sky Tools 3* at www.skyhound.com)

Chapter 2

The Best Viewing: Where, When, and How

Astronomers are known for their constant search for locations that have the best viewing, ranging from the remotest parts of Earth, such as the Atacama Desert in Chile, to Earth's orbit just above our atmosphere. This is true in our present time as much as it was in the eighteenth century. But at times, we simply have to make the best use of our circumstances and observe in places that are less than perfect.

In 1771 Charles Messier published the first part of his catalog listing 45 objects, and all of those observations were made in the city of Paris. Although there was some interference from the various types of candle lamps in use at the time, the city skies of his day were still far less polluted than ours. His skies were not drowning with the incandescent light of factories, stores, and electric street lights. Even so, it is still impressive that he was able to observe so many notable objects in his favorite instrument, the 3.5-in. (90 mm) achromatic refractor, set atop an observatory in the Hotel Cluny in Paris and also used at the Royal College of France. (Although it is true that he also used 7.5-in. (190 mm) and 8-in. (200 mm) reflectors because they were highly polished metal mirrors made from speculum, their equivalent aperture in today's terms would only be about a 3.5- to 4-in. reflector). It was in those locations where he lived and worked, making it convenient and almost a necessity to make whatever observations he could at or near those locations in the heart of Paris.

By 1784 up to 103 objects had been published in the catalog compiled by Messier and his associate Pierre Méchain. (Later entries, 104–110, were added since there is evidence that they had been previously identified by the duo, although not published.)

If you are beginning your journey into astronomical observing, keep in mind that Charles Messier was not born with a telescope in his hands, so to speak.

© Springer International Publishing Switzerland 2015
D. A. Jenkins, *First Light and Beyond*, The Patrick Moore Practical Astronomy Series,
DOI 10.1007/978-3-319-18851-5_2

Although today Messier is known as one of the most capable and popular astronomers in recorded history, in his childhood he had no formal training in astronomy.
It would not be until his adult years that he would gain the experience in observing
that we still benefit from in this, the twenty-first century.

Becoming a successful astronomical observer is definitely an achievement that
takes time, and benefits from continued and renewed learning. It is also not something that most of us can learn completely on our own. Of course, close study of
materials such as this book is essential, but along with that, association with other
observers can go a long way. Let's briefly consider how this was the case with
Messier.

Charles Messier was born on June 26, 1730, in the rural town of Badonviller,
located in the Lorraine region of France. At an early age his father died, leaving his
older brother Hyacinthe to take care of him and his five other siblings. Hyacinthe
taught him many skills that would aid in his possibly becoming a town administrator, as their father had been. When Messier was age 21 two important posts became
available, one in administration, the other as an assistant to Joseph Delisle, head of
astronomy at the Royal College of France. It was clear that he would benefit more
from the job with Delisle, and we today can be happy that Delisle agreed.

At first, Messier's work involved drawing geographic maps. However, by 1753
he was learning astronomy, even assisting Delisle in recording the transit of
Mercury. Impressed by Messier's skill and work ethic, it was in 1757 that Joseph
Delisle, by then Astronomer of the Navy, asked Messier to begin searching for the
comet predicted by Edmond Halley to return in 1758. History records that it was
German amateur astronomer Johann Georg Palitzsch who found it first on
December 25, 1758, although he thought it was simply a star with nebulosity. Then,
just 1 month later in January 1759, Charles Messier also made an independent
observation of the comet, which he had been expecting.

Despite working under the shadow of Joseph Delisle, Messier continued to make
his own observations and to diligently record them over the next 9 years. By the
year 1771 he had discovered eight comets, and published the first 45 objects of his
Catalog of the Nebulae and Star Clusters. Although he began his celestial journey
seeking comets and recording objects that could be mistaken for comets, it is clear
from his own comments in his catalog that he was moved to make a precise recording of the locations and descriptions of many deep sky objects. Messier was finally
recognized for his achievements first in 1770 by being elected to the Royal
Academy of Sciences, and then later in 1771 he was appointed Astronomer of the
Navy. (Delisle had died in 1768.)

At this point in his career, we might be inclined to think that he reached the pin-
nacle of his achievements, that in his position of Astronomer of the Navy he must
have known all there was to know in his time. But that was not the case—he had
many more discoveries forthcoming. By 1781 the catalog by Messier had increased
to 103 objects, and by 1801 he had discovered at least 15 comets!

Even though at the time he may have felt that his colleague William Herschel
was surpassing him because he had cataloged over 1500 objects by 1788, he could
still be proud that he was the first to take on such a detailed task. It was also his

work that inspired his contemporaries, along with astronomers of our time, to create catalogs. Today there are enough deep sky catalogs in existence that amateurs will discover that most of what they wish to observe has already been documented somewhere. Even so, Messier's catalog continues to be among the most used and cited astronomical references in our time.

There are really no limits to what you can achieve as an amateur astronomer. Some have observed thousands of objects over the years and have even viewed galaxies in amateur telescopes, previously thought by some professionals to be visible only in observatory-level instruments! There are also amateurs who, in a relatively short period of time, have learned to sketch with such accuracy and beauty that their work rivals photographs of certain celestial objects. Other recent achievements of amateur astronomers (including some who are children and teenagers) are discoveries of supernovae, exoplanets, and detailed analyses of variable and binary stars, to name just a few.

However, before we find an earth-like planet near Alpha Centauri or discover a supernovae in a nearby galaxy, like Charles Messier we need to start with assessing our location—where we can find the best viewing near our home, when we should observe, and how we can do it right. Our discussion will also include an understanding of the components of a stargazing forecast, along with many additional aspects of "when" to view certain celestial targets.

Where to View the Sky Close to Home

For those of us who do not live in a remote rural area, the question of where to find a decent view of the night sky close to home is very pertinent. Light pollution seems to have reached virtually everywhere. Most of us have seen NASA's images of the lights covering the globe and realize that truly dark skies are becoming a challenge to find (Fig. 2.1). Also, because of the dictates of our economy, many people find that they must live and work in or near a large city. Even suburban areas and semi-rural areas are still affected by the glow of not too distant city light. If you do not live in an area with dark skies, it is probably not convenient to travel to one every time you wish to observe. With a little perseverance and ingenuity several suitable places can be found to observe the night sky close to home.

First, your own backyard or terrace will be the most practical place for you to observe, because of convenience and the potential for regular use. Even if your residence suffers from local or distant light pollution, many targets can be viewed in urban and suburban areas. This includes of course the Moon, along with the brighter planets, and even many deep sky objects. Some unwanted light can be mitigated by using a light pollution filter. These are designed to filter out light from longer wavelengths that come from city lights while allowing the light from stars and nebulae to pass through the eyepiece. (In the next chapter, we will take a look at a few of these and compare their function so as to allow you to predict how they might perform with your telescope and viewing locations.) Your telescope can also

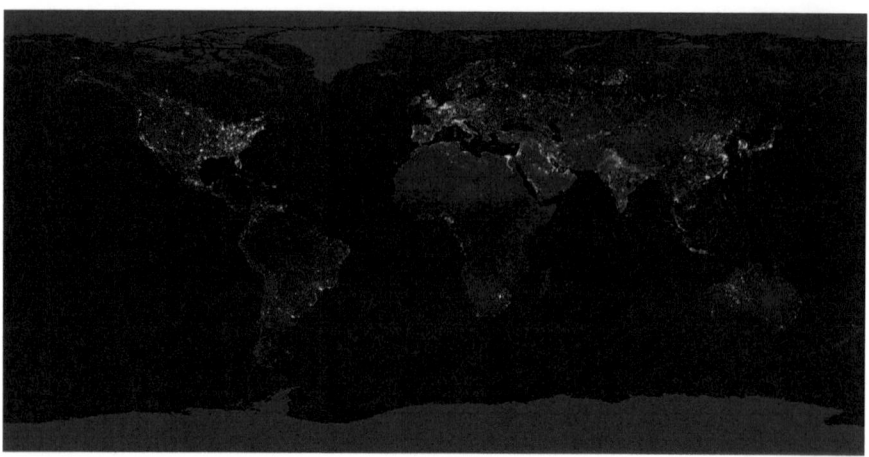

Fig. 2.1 Composite map of the world assembled from data acquired by the Suomi NPP satellite in April and October 2012. (Illustration courtesy of NASA Earth Observatory/NOAA NGDC)

be fitted with a light shield on the front end of the tube so as to limit stray light from entering in.

At times it is a street light or a light from another place nearby that interferes with your viewing. Some observers have reported constructing a simple light shield made from black cloth mounted on poles stuck into the ground or on whatever else you might find that is tall enough to block the incoming light. If you have a fence around your home, such a cloth could be easily mounted on it to provide shielding. Also examine the light fixtures around your home, so as to be sure that they do not intrude on your neighbors. They should be designed in such a way that when in use, the light points downward on the property as opposed to flooding unwanted light outward or upward, which is an inefficient use of energy. If not, appropriate shielding can be added to your light fixtures, some of them made from household materials.

During your observing time, you will want to be sure to limit, if not eliminate altogether, any lights from inside your house so it will not impede your view of the sky. Every little bit helps, and this may involve some coordinating with family or roommates who may wish to use indoor lighting at the same time.

In the case of unwanted light coming from a neighbor, you could simply speak to your neighbor in a polite manner about how the light is impacting you. A more positive approach may be to share some sights with him or her through your telescope. This provides an easy way to open up a discussion about the effects of light pollution on viewing, and your neighbor is likely to suggest modifying the bothersome light before you even directly ask. Additionally, you will be sharing a little corner of our universe with someone who may end up becoming just as interested in dark sky friendly lighting and astronomy as you are.

For light coming from a street light that is not properly shielded, the task may require a little more persistence if there is not a way of shielding the incoming light from your property. Speaking to the local municipality or city council at public hearings about your concern for environmentally appropriate lighting would be the first step. Many of them are open to public concerns about light pollution. After all, it is illegal to litter the streets, and hopefully through more education about the issue and persistence of the community, it will be illegal to litter our night skies with excessive light.

When it comes to lighting, no doubt many of us would love it if at night time there were no lights at all to contend with, although realistically we do have to recognize that some lighting is necessary. Be sure to communicate to others that light at night itself is not bad, and does not need to be eliminated entirely. Often it is just the wrong kind of lighting that is the problem.

For example, light should help us see the roadways better, or help to provide a visible path to an entryway. However, most street lights in cities around the world are exposed on each side, sending beams of wasted and unwanted light everywhere, including into our beautiful skies. In the United States alone, more than $10 billion per year is spent on light that is improperly directed and thus wasted by shining into the sky, according to the International Dark-Sky Association. That also means that thousands of tons of carbon dioxide is needlessly being dumped into the atmosphere each year as well. In view of this plight what is needed is more investment in fixtures that point light down towards the true target and that are shielded so that stray light does not pollute the night sky. There is a lot that is being done to help ensure that this concern is recognized and continues to be a priority so that we do not lose this precious possession that is inherent to our planet. It sounds so simple—we have day and then we have night, but humans have become good at destroying the night. What is being done about this issue? More details will be forthcoming on this in Chap. 6, but in the meantime let's return to our discussion on where to find the best views of the heavens close to home.

Many cities today are building more public parks for people to enjoy in the daytime or in some cases also at night. Depending on the location, it is possible that a city or residential park may provide you with a darker or less obstructed view than your home. Additionally, a public school yard could be a possible place to set up your telescope. Permission should always be requested from a school official, however, as in many places to do otherwise would be considered trespassing. Be sure to check out what is stipulated by local law, school officials, and law enforcement prior to entering the property. When asking permission from a school official, it would also make sense to offer views of the heavens as part of an outreach program for the children at school; then there would be a benefit to the school in allowing occasional use of the property. Another option would be to check out local campgrounds. Although there could be some distractions from patrons engaging in more boisterous activities, usually there are areas where you can camp that are quiet and ideal for a view of the sky. Let us also not forget the obvious, that a friend or relative living nearby may let you use their property at times and may welcome the opportunity to look through your telescope.

All across the world, there exists a growing network of people that are members of an astronomy club. An astronomy club has already done much of the legwork that we have described in locating and securing permission for use of appropriate places to observe the night sky. Some clubs also offer additional benefits, including discounts on lodging or subscriptions to astronomy magazines. Most also offer public viewing nights where you can casually attend and get acquainted with a few people and take a look at the heavens. Often, clubs will have multiple locations that members can use for observing, at least one that tends to be very close to many members, and then one that is considered to be their best dark sky site that may be a little further away. This provides camaraderie along with the security of having observing partners when in more secluded areas. Take for example a club in the United States in Phoenix, Arizona, called the Saguaro Astronomy Club (www.saguaroastro.org). According to its website it was established in 1977 and currently has over 100 members. In addition to a lecture and a group observing session each month, they also have subgroups within the club that focus on specific interests of members, whether it is the concerns of beginners (the Novice Group) or a group dedicated to finding very faint objects (the Deep-Sky Group). Membership is usually granted for a small annual fee that helps cover the cost of various benefits.

Consider another club located in Chipping Norton, in Oxfordshire, England, called the Chipping North Amateur Astronomy Group (www.cnaag.com). This club has grown from just three members to about 40 at the present time. Its members have access to several observing sites that provide dark skies. The chairman of the club offers a monthly astronomy radio show, and each month the club also publishes an informative astronomy newsletter. The club welcomes people no matter what their prior experience level is and provides a lot of outreach astronomy sessions for local groups and schoolchildren.

These two examples are just a sample of what you can find in your local astronomy clubs, and well represent what most have to offer at the very least. You will no doubt find a level of involvement in an astronomy club that will fit your personal circumstances and will not entail any pressure. After all, it's all about people getting together to share and to help each other enjoy one of the awesome beauties of nature—a star studded sky!

When Is the Best Time to Observe?

The answer to this question depends greatly on the type of object being observed, what the weather and other stargazing conditions are like, and the time certain objects rise and set. We will take into account what Universal Time is, and how it affects the times you choose to view certain objects. The first consideration, however, is what is the best time for you personally to observe. When will you have the most energy? How will the times you choose be affected by your work schedule or family life?

Some people prefer to observe in the early evening before they retire to bed. Others are real night owls and observe late at night into the early morning hours.

If so, it may be possible to get in an hour or so of observing before dawn. For those that work in the afternoon into the early evening, some may choose to observe before going to bed, which may be easier than rising earlier than usual. All of this really depends on personal preference and schedules, but once you find what works best, you'll no doubt find yourself wanting to extend the length of your observing sessions more and more. In any case, a nap before you observe is wise, so as to help you stay alert to your surroundings, and rest will help your eyes to see more details when you view objects at night. A tired mind or body will only lead to exhaustion and missed viewing opportunities of objects that may have been on your list to observe.

What Is Universal Time?

In the late nineteenth century, astronomers agreed to use a common reference of time that everyone would understand. This would provide a simple means of expressing exactly when a celestial object was discovered or when a celestial event occurred. The time zone that was agreed upon was that of Greenwich, England, and was first referred to as Greenwich Mean Time. For astronomers, this is now most commonly referred to as Universal Time, and is expressed in a format similar to military time. For example, 12:00 a.m. local time would be 0:00 Universal Time, 1:00 a.m. local time would be 01:00 Universal Time, 2:00 p.m. becomes 14:00 UT, and so on. Sometimes, therefore, the best time to observe is when an event happens at Universal Time, if it will be visible in your part of the world.

What do we mean by that? Well, it is important to note that to convert UT to your local time, you will need to adjust it according to your time zone, which may affect the date also. This in turn may also affect whether or not an event is partially visible in your location, or in some case not at all. Consider for example a conjunction that occurred on August 18, 2014, when Venus was just 0.2° north of Jupiter (with M44, the Beehive Cluster, also nearby), that happened at about 05:00 UT. However if you lived, let's say, on the western coast of the United States, then that time would have been 9:00 p.m. local time on August 17, 2014. At that time on the west coast of the United States, the planets had not yet rose, but by dawn local time they would still be seen as fairly close to one another, about 0.5° apart. The chart below should help you to convert UT to local time for several time zones in North America.

During Standard Time

Eastern Standard Time:	Subtract	5 h
Central Standard Time:	Subtract	6 h
Mountain Standard Time:	Subtract	7 h
Pacific Standard Time:	Subtract	8 h
Alaska Standard Time:	Subtract	9 h
Hawaii Standard Time:	Subtract	10 h

During Daylight Savings Time

Eastern Standard Time:	Subtract	4 h
Central Standard Time:	Subtract	5 h
Mountain Standard Time:	Subtract	6 h
Pacific Standard Time:	Subtract	7 h
Alaska Standard Time:	Subtract	8 h
Hawaii Standard Time:	Subtract	9 h

When to View the Planets

From rising to setting, the planets follow a clear apparent path through the sky along what is called the ecliptic. Generally speaking, as the planet you seek rises higher in the sky along the ecliptic, the better the view will be, since it will gain altitude above much of the interference from light and haze near the horizon. Times for planetary viewing are among the most flexible, since the planets are so much brighter than stars and thus much easier to locate. It does not require extremely dark skies, and even if the Moon is out, this need not be a reason to delay most planetary observations. One of the most fascinating times to observe a planet is when it occulted by the Moon or passes near it during conjunction (see Fig. 2.2). Current dates for such events can be found in annually published almanacs such as *Sky & Telescope's Stargazer's Almanac* or the *Observer's Handbook* published by the Royal Astronomical Society of Canada. Another way to keep abreast of these periodic happenings is to check the monthly *Sky at a Glance* section in the print or online edition of *Sky & Telescope* magazine.

Another advantage a planetary observer has is being able to view the planets even when there is some cloud cover. Usually cloud cover spells disaster for the deep sky observer, as these objects require very clear conditions to reveal much detail, as well as the fact that a cloudy forecast often indicates that the objects may be covered up altogether. However, the bold light of a planetary disk is not so delicate so as to preclude viewing during some scattered cloud conditions. Of course the optimum weather conditions for any astronomical observing would be completely clear skies.

When to Best View Meteor Showers

A meteor shower occurs as Earth moves through the path of debris from a comet; those particles left over from that comet are seen burning up in our atmosphere, in the form of what we call a meteor shower. This is different from a meteorite, which is a rocky object that actually hits Earth. Unlike meteors, meteorites come from asteroids, although some are believed to have come from the Moon and also Mars.

Fig. 2.2 Photo of the waxing crescent Moon in conjunction with Jupiter (*bottom*) and Venus (*top*), over Tokyo, Japan, March 26, 2012. (Photo courtesy of ESO/B. Tafreshi [twanight.org])

Several meteor showers can be seen throughout the year for about a night or two. Some average a nighttime show of 10 or less meteors per hour, while others can dazzle us with 100 or more per hour.

Meteors in a particular group appear to come from one region (constellation) of the sky, and astronomers refer to this point as the radiant. Meteor showers are named after the constellation containing this radiant point from which these beautiful streaks of light seem to originate. The Leonids come from the comet Tempel-Tuttle, and appear to radiate annually from the constellation Leo on about November 17. The Perseids come from comet Swift-Tuttle, and appear to radiate from the constellation Perseus on approximately August 12 of each year.

Meteor showers are a regular occurrence through each year and can provide enjoyable casual viewing as you observe them with the naked eye. Check your local monthly astronomy magazine for more information about upcoming showers or check online. The app *Meteor Shower Calendar* is free and also contains an easy to read format list of upcoming meteor showers with several customizable options. Although a specific time may be listed for the shower, keep in mind that most last for several days, and you may try observing a day or two before and after in addition to the specific date given. If you do not see very many at first do not give up! Often, the rate of meteors per hour increases as the radiant point rises higher in the night sky.

Fig. 2.3 The Milky Way. (Image courtesy of ESO/B. Tafreshi [twanight.org])

Another grand sight to view with just your naked eye or with binoculars is the cloudy band of the Milky Way, which is made up of a dense region of stars within our own galaxy. The Milky Way has become one of our lost commodities, no longer within the reach of our light-polluted cities. Although best seen from truly dark, remote sites, thankfully it can still be seen in some moderately light-polluted skies. During the summer months you will see the long arc of the Milky Way stretch from the northern sky all the way to the southern sky. In the winter months it will stretch between the eastern and western skies. Many of the objects in the Messier catalog can be found within the portions of the sky in or near the Milky Way, many of which are visible to the naked eye. On a cloudy night you may enjoy viewing its seemingly endless swaths on the Spitzer Space Telescope website: http://www.spitzer.caltech.edu/glimpse360. Even so, there is still nothing like being outside underneath it all, soaking it in with your own eyes (Fig. 2.3).

If you live in far northern latitudes the aurora borealis (also sometimes referred to as northern lights) is a dynamic sight to observe. Auroral displays are caused by an influx of electrons and protons coming from the Sun, propelled by the solar wind, into Earth's magnetosphere. Although more displays are expected during years of peak solar activity, they can occur at any time throughout the year. These charged particles tend to stay in regions near the poles, either north or south (in the southern hemisphere it is called the aurora australis or southern lights). In the northern hemisphere, regions most likely to see the display would be in Greenland, Alaska, and others near those latitudes. However, auroras have been reported as far

Fig. 2.4 The Aurora Borealis is seen in the night sky in Anchorage, Alaska, on March 8, 2012. (Photo by Justin Connaher, provided courtesy of NASA)

south as latitude 40° north. They take many different shapes such as arcs, bands, and patches, appearing to drop down from the zenith overhead to the ground in sheets and then disappear as rapidly as they began.

Auroras also can be seen in varying colors, depending on the level of energy of the particles that are caught in Earth's magnetic fields. If the incoming energy is low we tend to see more red colors, and green or blue if the energy is high. One observer near the west coast in the state of Washington in the United States reported seeing what appeared to be a large curtain of red hues descending over the roadway while driving home one night in 2006. Many observers have photographed these beautiful displays on Earth, and the Hubble Telescope has also recorded aurorae on Jupiter and Saturn. Keep your eyes open—you never know where you may see the aurora borealis next (Fig. 2.4).

When to View Deep Sky Objects

The short answer to the question of when to best observe deep sky objects would be whenever you can! If you were to ask a group of avid deep sky observers however, the answers would probably vary. Here are a few tips that will help you select the optimum times for viewing.

- Seeing detail in nebulae, star clusters, and galaxies requires a reasonably dark sky. For the best visual experiences, observe well after evening twilight, when these objects look their best against the background of a truly dark sky.

- Deep sky phenomena can be observed anywhere your telescope can point, but the most pleasing views will be when the object is well above the horizon, away from haze and light pollution from nearby cities.
- Observe when your telescope has cooled down to the surrounding air temperature. When you first set up your 'scope, it will usually be much warmer than the air outside. This causes tube currents from the heat that is given off and can wreak havoc with the images you see. An easy solution is to set up your telescope outside in advance of an observing session, allowing it to cool down as well as giving your eyes plenty of time to adjust to the dark.
- Try to avoid observing when you are near warm structures such as cement, buildings, and cars. These give off heat, causing much of the same trouble with viewing images as do tube currents. As we will see shortly, there are plenty of other atmospheric conditions to be aware of that are beyond our control, so limiting those that we can as much as possible is important.
- It is not true that you can only view deep sky objects on the night of the new Moon. Since deep sky phenomena are best seen under the darkest conditions, it is best if the Moon is not visible while you observe them. The night of the new Moon does provide that condition, but it is not the only time. The day or two following the new Moon are also very good times for these observations since the Moon will only be a thin crescent. Even better, during several days before a new Moon, the Moon does not rise until very late at night, sometimes even after midnight. This allows several hours of dark for deep sky observations before the Moon causes interference. You can check Moon rise and Moon set times on the Internet or check an annual almanac such as was mentioned earlier. Make use of every possible hour of the moonless sky that is available!

Another important aspect of optimal viewing times for deep sky objects involves sidereal time. The sidereal time of an object is the point when it is located at its zenith, or highest point overhead. This moment is also called *culmination*, and this point for various deep sky objects is often referenced in almanacs, magazine articles, and observing guides. The easiest way to find out, however, is to simply look at a planisphere, set it to the day and time you are observing, and look at what constellations are along the meridian overhead. Those constellations are well placed far away from the haze of the horizon, and will usually provide a better view of the deep sky objects contained within their boundaries.

Observing Conditions

Before observing begins, amateur astronomers will often check and discuss the local stargazing weather conditions. There are four main observing conditions that concern us the most: cloud cover, darkness, seeing, and transparency.

Cloud cover is the most obvious of these four aspects. Although a little cloud cover does not always ruin an observing session, it is still undesirable. It is reasonable to even expect some patches of clouds, as oftentimes they linger near one or more of the horizons. Sometimes a few stray clouds may float across the sky, and this is normal. It becomes a problem if there are many clouds in at least one or more directions. If you are trying to gauge whether or not to set up your telescope, sometimes it is simply a matter of waiting until twilight to see how the sky turns out. Many amateur astronomers have seen countless clouds cover the sky during the day, but by evening have watched them disappear just in time for some great observing. Of course if there is a 30 % chance or more of rain in your local forecast, chances are high that the clouds will not abate by nightfall.

The measurement of *darkness* relates to the time of day at which the sky has become truly dark, generally an hour or so after twilight. However darkness can also be affected by the existence of light pollution and the presence of the Moon. In times past it has often been measured simply by determining the magnitude of the faintest star that can be observed at a location on a particular night. However in 2001 John Bortle published an article in the February 2001 issue of *Sky & Telescope* magazine revealing the criteria for his scale to measure the true darkness of any given observing location. His attention to detail really covers everything that affects the quality of a dark sky. You can read his article re-printed on the magazine's website (www.skyandtelescope.com). The Bortle Scale, as it is now known, is widely used in places all over the world. In fact each location designated by the International Dark-Sky Association as either a Dark Sky Reserve or a Dark Sky Park has been given a rating on the scale of 1–9, with 1 being the most pristine skies available. Several locations have been so designated throughout the United States, the United Kingdom, and in Europe. Some state parks and astronomy clubs have now adopted use of the Bortle Scale so that visitors can readily see and understand how their location rates in true darkness.

The measurement of *seeing* is an indication of the amount of turbulence found in the air. When we look up into the night sky, the light we see from the stars has to travel through hundreds of miles of the Earth's atmosphere. The air limits the perceived magnitude of the stars and also causes some degradation in the image quality. (That is why the Hubble Telescope has such an advantage as it views celestial objects from an orbit of 353 miles [569 km] above our atmosphere, free of its adverse effects.) If the quality of the air is very steady then the seeing conditions can be described as good.

In addition to bad air turbulence, varying air temperatures also affect the quality of the seeing on a given night. That is why it is not good to view the heavens over nearby buildings that may be giving off heat absorbed throughout the day. Imagine looking at corals through the surface of a body of water—your view would be interrupted by a ripple effect. This is similar to what happens when viewing objects such as planets under poor seeing conditions, and this effect is especially apparent when viewing with higher magnifications.

One way to test the quality of seeing at the beginning of an observing session is to view a planet first at high power and take a look at how steady the image comes through. Another object to test would be a star situated far above the horizon; note how much it seems to twinkle. Some twinkling in the image is unavoidable, but if it is so much that the image of the star seems to break apart, then you are definitely going to be hampered by poor seeing for some time. As the night moves on and the air cools down a bit, the seeing may become more steady. Even under the best see-ing conditions, however, it is generally not recommended to try and use magnifica-tion any higher than 50× per inch of aperture. Even owners of larger-sized telescopes of 10 in. or more (254 mm) rarely use more than 300–400×.

Whether or not skies are *transparent* is an indication of how much moisture there is in the air, along with other trespassers such as aerosols (e.g., pollen, ash, and smoke). As is the case with the aspect of darkness, local light pollution and the pres-ence of dust particles will also affect transparency. For example in the southwestern United States, oftentimes there are windy conditions that stir up a lot of dust parti-cles. It is common for observers to find after being outside all night that their tele-scope optics have accumulated almost enough dust to warrant a cleaning. That is the kind of dust that will impact the level of transparency in the sky, and thus limit how much you will be able to see. Warm humid days also produce nights that are not very transparent. On the other hand, the air on a winter day after a cold front has gone through can be much drier and cleaner because dust particles have been cleared out, causing a transparent sky to prevail. These are the kinds of conditions under which faint deep sky objects are best viewed. Usually, however, astronomers end up with a mixture of some aspects being good while others may be just average. On those few occasions when all aspects of the stargazing forecast are excellent, count your-self very fortunate and make the most of such beautiful observing conditions.

All of the attributes that have been discussed here, and more, can be found on the Clear Sky Chart website (www.cleardarksky.com), which has stargazing fore-casts available for many cities in North America. Simply select your city or one nearby and begin planning the timing of your observing sessions. For specific hours during each day, a forecast is given for cloud cover, transparency, seeing, darkness, wind, humidity, and temperature. Another way to obtain a forecast is through an app called *Scope Nights: Astronomy Weather Reports*. It covers the United States, the United Kingdom, and other locations worldwide with updates that range from a 3-h forecast up to ten nights out. It predicts cloud cover, humidity, wind, tempera-ture, and displays the current Moon phase.

Remember, though, that a forecast is only a probable prediction, and that condi-tions can quickly change. With a little patience and time you might just beat an unfavorable forecast.

How to Have the Best Viewing Despite Obstacles

There is rarely such a thing as a perfect observing session, but with careful planning you can always have outstanding ones. Obstacles will often arise, and their negative effects can be mitigated if you know how to deal with them. Let's take a look at a few of these obstacles.

First, fatigue. As touched upon earlier, fatigue comes when we try to observe when we're already too tired. This can lead to confusion and irritability. The result is the inability to find the tiny targets you're looking for, and also makes it very difficult to absorb new information from fellow club members or your observing partner. The best views in the eyepiece come when you are alert but relaxed.

Related to this point of being relaxed is a practical suggestion to bring a chair to your observing session so that you can rest a little. It can also be used to keep you seated while observing some objects that are not near the meridian. This will help you preserve valuable energy that can be better used in viewing more details at the eyepiece. Here is a suggestion on how to stay seated for almost an entire evening. Try starting your survey low in the western skies, beginning with objects that are just about to set. Then work your way a little bit east, targeting a few new objects a bit higher in the sky. When you have reached the highest point that you can comfortably observe while seated, slowly let objects come to you as they move westward. Remember you can also move your 'scope anywhere along a line of varying degrees of declination (latitude running from north to south), investigating many objects, and your 'scope will still remain at the right ascension or altitude (longitude) that you want.

Have you already had an observing session that lasted for many hours? Then you may be experiencing a tired eye that came from squinting your unused eye at the eyepiece. There are two ways around this: purchase a binocular viewer (also called a binoviewer) or use an inexpensive eye patch. The eye patch allows you to keep your eye open while using the other eye, making you just a little more relaxed and able to get that much more out of your observing.

Another obstacle with which to contend is the presence of insects. A night spent swatting and scratching after a lot of insect bites can put a damper on the fun of observing, so come prepared. Bring your preferred insect repellent—whether it's the standard variety containing Deet or one of the natural sprays that have become popular. When it comes to observing in fields or other similar areas more open to nature, you'll probably find that the critters generally keep to themselves. However, if it becomes a matter of concern, try an observing area that is fenced in or otherwise cordoned off somehow. Others take their dog with them to remote observing sites because there have been isolated cases of wild animals approaching. While those cases are rare, many amateurs have been able to feel much more secure and comfortable with a dog or two around to ward off any unwanted visitors. Of course if you do decide to bring a pet to a star party, be sure to check the local star party etiquette that doing so does not disturb others.

Lack of familiarity with your equipment is another obstacle that can be overcome with good planning. Even if your telescope breaks down into a few simple pieces, you will probably have your car packed with quite a few things when going out to an observing site. This may include star charts, chairs, binoculars, a red light, snacks, and more. It would be very good idea to practice setting up your equipment and taking it down again the day before going out to a site. There is nothing worse than trying to figure out everything in the dark for the first time. Be aware of any small parts or pieces of equipment that are easy to lose. Put them in a large plastic sealable bag or a Ziplock bag to prevent them from getting lost. Bring along spare batteries with you for any electrical equipment that you use. Also, try to have all of your equipment packed in your vehicle well ahead of time.

Everyone prefers to begin observing at different times, and with good red light-ing available, setup can be done even in the dark. However, if at all possible, espe-cially if your equipment is still relatively new to you, try to begin setup while there is still some daylight left, perhaps an hour or so before twilight. This gives you plenty of time and allows you to get through any unforeseen snags or delays that could come up during your setup time. Collimation and alignment of your finder-scope and/or red dot finder will be more easily accomplished with a little daylight on your side than in the complete dark. By the time twilight comes and the first stars appear, you will be ready to fine tune the alignment of your finder to the main 'scope by centering a star.

Now comes the moment when you look through the 'scope and wonder if it is collimated properly and if it is showing you the best views it is capable of provid-ing. Is that star supposed to look like that? Why does the image seem so hazy? These are the things that you will have to sort through. The doubts you have about the initial view in your telescope often come from either it not being properly col-limated or because you are experiencing adverse observing conditions of some kind. To check how the seeing is in your sky, turn your telescope to one of the planets, if visible. Does the image show a clean steady disc or does it appear turbulent? An alternative would be to view a bright star. Does the image come into focus easily, giving a clear sharp view of the star or is the view hazy with excessive twinkle? All stars twinkle because of atmospheric conditions, but when the seeing is poor, this effect is intensified. To get an idea of just how dark and transparent your sky is on a particular night, try turning to a more challenging object to test your observing conditions. Here is a sample list of four targets you might use, one for each season of the year:

Winter:	The Trapezium of four stars located inside the Orion Nebula (M42)
Spring:	The Sombrero Galaxy (M104)
Summer:	The Trifid Nebula (M20)
Fall:	The Andromeda Galaxy (M31)

If you can see the details in one of these objects, chances are your local weather conditions will favor you with a good night of observing. Let's briefly look at what those details are in each of the above examples. First, the Trapezium inside the heart of the Orion Nebula. In a 6-in. (152 mm) you should clearly see the four individual stars shaped in the form of a trapezoid. If conditions are excellent the stars should show pinpoint clarity. Second, try to see the beautiful dust lane in the Sombrero Galaxy. In poor conditions it will be hazy and unclear. Under ideal condi-tions the dark dust lane should appear sharp.

Third, in the Trifid Nebula, you are looking for the dust lanes that appear to divide this nebula into separate parts. Under the very best conditions you should be able to make out three dust lanes (although in photographs four are visible). Fourth, the Andromeda Galaxy provides a good test. If conditions are good you should be able to see this galaxy with your naked eye. If the transparency or seeing is so poor that you cannot see it, that is a definite indicator that conditions are not good. Additionally, any planet that is visible would also be a great way to test the seeing. If a planet appears blurry, as if there is a sheet of plastic being waved across its disk, there is poor seeing.

If you have any trouble finding an appropriate object with which to test your observing conditions, just ask for help from someone more experienced. Or if you are using a computerized telescope you can type in one of the objects mentioned above in the hand controller and have a look. Otherwise, by the time you complete the celestial treks in Chaps. 4, 5, and 6, you should be fully capable of finding many targets to use in testing your observing conditions as well as all of the ones mentioned above. By taking a few minutes to test conditions at the start of your session you will have a better idea of what to expect, be able to recognize any improvement in the viewing as the night goes on, and begin training your eye to make better observations from the outset. Try returning to one of the objects you used as a test after about an hour or so. You may find conditions have slightly improved, and regardless, you are sure to discover more detail with patience and a second look.

Helpful Things to Consider

How will the local stargazing conditions affect what I observe tonight? Have I done my best to eliminate nearby sources of stray light from my observing area? Will there be any special astronomical events to observe during my next observing session? What obstacles may interfere with my next sessions, and how can I plan ahead to limit their possible effects?

Here are a few reminders and additional points to keep in mind:

- Remember, sometimes it is not possible to observe under pristine dark skies, so make the best use of your local circumstances. Look for uncommon corners in your area that may provide reasonably good views of the sky. Be sure to seek permission before setting up your equipment.
- Check with local school districts and college campuses, as they may already have an observing area that is open to the public on certain nights, and they may let you use it at other times as well.
- Local clubs often feature community outreach nights where everyone brings a telescope to share views of the sky. Many of these locations are close to suburban or urban areas and provide some measure of relief from glaring light pollution sources.
- Keep informed on local astronomical events each week. This allows you to maximize your allotted observing time so that it includes not only the objects that you can always view during a certain month, but also those special events such as conjunctions, meteor showers, and occultations.

- Messier's catalog began as a list of objects not to be confused with comets, but then turned into something much more grand—a treasure trove of deep sky objects. So you also should plan a little extra observing time for encountering phenomena that you did not expect.

No matter what your local observing circumstances are, never let them prevent you from having a good experience. Even Charles Messier had to learn astronomy, step by step, until he finally became one of the most successful astronomers to turn a telescope to the night sky. He also did so with less than perfect observing equipment and locations. With diligence, even below average conditions can be turned around for the better and be greatly improved. You may not discover an exoplanet on your next night out, but with patience it can be a successful one. Beautiful celestial sights at the eyepiece are very rewarding, but they will not appear without your first knowing where, when, and how to obtain them. With a little bit of forethought and planning, you will always be able to find the best viewing.

Chapter 3

Equipping Yourself
for the View

From a very young age Edward Emerson Barnard had a passion for looking at the stars with the naked eye, a real pleasure in an era that had essentially no light pollution. He was raised in a very modest household during post-Civil War days, and so at age nine was required to work in order to help support his family. This young man from Nashville, Tennessee, would become known as one of the greatest astronomers of the late nineteenth and early twentieth centuries.

E. E. Barnard faithfully worked at a photograph gallery for 15 years acquiring experience and nurturing friendships that would serve him well in later years, both personally and professionally. As fate would have it, one day in 1876 he chanced upon a collection of astronomical writings by Thomas Dick that profoundly increased his interest in astronomy. When he was about 19 years old, Barnard's friend and co-worker, James Braid, helped him construct his first telescope. It was only about 1 in. in diameter and very crude. Soon afterwards Barnard would go through at least three other telescopes, none of them being the right equipment to suit his purpose of scanning the vast stars of the Milky Way or exploring the planets in greater detail. Then in 1877 he heard about an optician in New York City who was selling a telescope. Barnard had finally found the equipment he was looking for, a well-made 5-in. refractor on an equatorial mount. Although living on a modest salary of $12 per week, he spent $380 on his new prized possession. It was with this telescope that he would make many of his initial discoveries that would propel him into astronomical history. His striking contributions to astronomy included the discovery of the fifth moon of Jupiter (Amalthea), several comets, dark nebulae, and much more.

In our twenty-first century, when it comes to telescopes and their accessories, we have many more choices than anyone in Barnard's day could have imagined.

© Springer International Publishing Switzerland 2015

D. A. Jenkins, *First Light and Beyond*, The Patrick Moore Practical Astronomy Series, DOI 10.1007/978-3-319-18851-5_3

Even so, at first glance it could be easy for you to believe that you have all the right equipment and will never need anything else. You own a telescope, an eyepiece or two, so what else could be needed? In that case you might be severely limiting your potential given the rich developments in amateur astronomy equipment. On the other side of the spectrum, you may have become convinced that being a successful astronomical observer means owning a computerized telescope, with the GOTO feature along with all of the other bells and whistles that come with it. In reality it would be wise to have a moderate view. You can expect that some additional expenditure will be required at some point. However, you do not need to spend an exorbitant amount of money for updated observing equipment. By now you have no doubt guessed there are several things that could potentially be of great benefit. Taking this idea just a bit further, consider that equipping yourself for the view is not limited to things that can be bought. Someone who wants to climb Mount Everest may have the best equipment money can buy, but the trip may not be successful if that person does not have endurance, the right attitude, and the right habits. So then, in addition to astronomical equipment we will also take a look at skills, attitudes, and habits that are sure to contribute to your success in visual observing.

We can categorize our review into four areas:

1. Equipment that you should try to obtain in the near future.
2. Items that are not urgent to acquire.
3. Useful household items you will need and likely already own.
4. Skills, attitudes, and habits that should be developed.

Keep in mind that needs and wants mean different things to different people. When it comes to equipment, for some people certain items must be obtained, depending on what areas of observing especially interest them. For other observers, those same items may not be important at all. The recommended items in the first category should be considered to be tools that you really must obtain. However in the second category, items that are not so urgent, you may find some items listed that you feel are definitely required. You will develop your own personality as an amateur astronomer, and ultimately you will decide the things that will best work for your observing sessions.

Equipment That You Should Obtain

Binoculars

Most telescope owners already have a good pair of binoculars, but if you do not, it is definitely something that you should acquire. The more detail you learn to see with binoculars, the better off you will be as an observer. With them you can learn to spot most if not all of the Messier objects. As you learn to recognize the star patterns that appear as seen in binoculars, this will help you to more quickly locate objects in your telescope. They really are a great compliment to the telescope, allowing you to easily see what is observable in the night sky from the perspective of a very wide field of view. In fact, at star parties it is common to see at least one person with binoculars mounted on a tripod, which for some is the ultimate way to view the sky—with both eyes. Hand held binoculars also offer an extremely comfortable way to observe when you are reclined in a chair. It can be great fun to trade views between telescope and binoculars on an observing night with a friend. You may also find that after a night of telescopic observing, it is nice to lean back and relax for just a few more minutes with binoculars, concluding the night with a larger view of what the sky had to offer.

A basic pair of binoculars is very inexpensive, (well under $100 or £65), although some can be priced considerably higher. This depends on the quality, aperture, and whether or not they are mounted. Unless you have very specific needs an inexpensive pair will do. In fact most people may already have a pair lying around the house somewhere, and most would do just fine for astronomical observing.

If you need to buy a pair of binoculars, try to get the best of both worlds—as much aperture as you can, with a reasonable magnification. Balancing both factors will give you an effective but very portable pair of binoculars that will serve the purpose of making sweeping views of the stars easy. A pair of 7×35 mm binoculars is suitable, but for not much more in cost, a little more light-gathering ability could be obtained with a configuration such as 8×40 mm, or even 7×50 mm, which is also common. With larger diameters, there often is an increase in magnification and weight. Both of these factors causes a problem with stability.

Try using a pair that magnifies 7× and compare that to one that magnifies 10×. You will notice that the higher the magnification, the harder it is to keep a steady view. If you go up to 15× or more, forget it. It's almost as if it reveals tremors in the hand that were previously unnoticed. A similar problem exists if the binoculars have an aperture of 60–70 mm or more; the heavier weight is just too much to hold a steady image over any reasonable length of time. One solution is to always remain seated while observing with arms placed on the armrests. Another is to buy binoculars with image stabilization, and of course this raises the cost considerably and often does not eliminate the problem completely. A third option is to mount the binoculars (generally anything with a 70 mm diameter or more) with one of many commercially made products. Over the years, there have been several suggested designs of mounts that can be constructed at home. For ideas on homemade mounts, check your favorite astronomy magazine's website.

Some binoculars are also waterproof, and feature focus adjustment for each eyepiece, which is great since our vision usually differs in each eye. Most binoculars come with anti-reflective coatings on the lenses to help prevent loss of light transmission. A few recommended manufacturers of binoculars that you may want to examine are Orion, Celestron, Bushnell, and Oberwerk. Each of them have high quality products within a wide range of prices. Features promoted by manufacturers beyond the ones already mentioned will make the cost go up considerably. The main factors to strive for here would be portability, light weight, aperture (as much as possible without becoming too heavy), and medium magnification (8–10 power).

A little higher power is nice because it will help you pick up more details, but just remember that going above 10× will increase the difficulty in maintaining a steady view.

Finderscopes

Unlike handheld binoculars, a magnifying finderscope that is well mounted to your telescope will be rock solid and not subject to an unsteady image due to hand tremors. Ideally it should have a quick release mechanism for easy removal and storage of the main telescope tube, although this is not an absolute requirement. Of course, the smaller finders should be avoided (under 40 mm in size), since they would not provide you with a wide enough field of view to locate your target for centering in the main telescope. If you have a magnifying finderscope, or need to replace an inadequate one, try to come as close as possible to the magnification and size of your binoculars. You will benefit greatly by being able to use your binoculars, and then see exactly or at least close to the same view in your finderscope. That way there is no confusion on what you see from one instrument to the next. For example, if you are looking at approximately a 5° field of view area surrounding the Andromeda Galaxy (M31) in binoculars, but then try using a small finder that gives perhaps half that field of view, it will be much harder to center your target for viewing in the telescope. A finder with a wider field of view also makes it easier to star hop.

Many of today's finders offer a correct image view just as you see in binoculars. Some feel that this is not a good thing because the view is not what you would see in your telescope. However, if we look at it from the perspective of matching what you see with your eyes and in binoculars, then it makes perfect sense. Look at the area near the Andromeda Galaxy in Fig. 3.1.

A bright star nearby is Mirach (Beta [β] Andromedae), which is in the neighboring constellation of Pegasus, so let's say you start there using binoculars. If you are using a correct image finder with a similar magnification and size as your binoculars, you will see very close to the same star field in your field of view, with no directional confusion. It would be very difficult to memorize the star field you saw in binoculars and then try to match the location in a smaller field of view, in addition to being flipped upside down. Instead, a correct image finder, if of similar proportions as your primary pair of binoculars, allows you to continue on seamlessly, as if you were still using the binoculars. You will need to get used to sighting down your tube towards your target since it is more challenging to do so looking straight down through the finder instead of up towards the sky. Some amateurs have both types of finders attached to their telescope, a reflex sight for ease of targeting and also a finder to match what they observe in binoculars. Once your field of view matches what you see in binoculars, you can then continue to star hop on to your target, and your telescope will follow your every move because it is aligned with the finderscope.

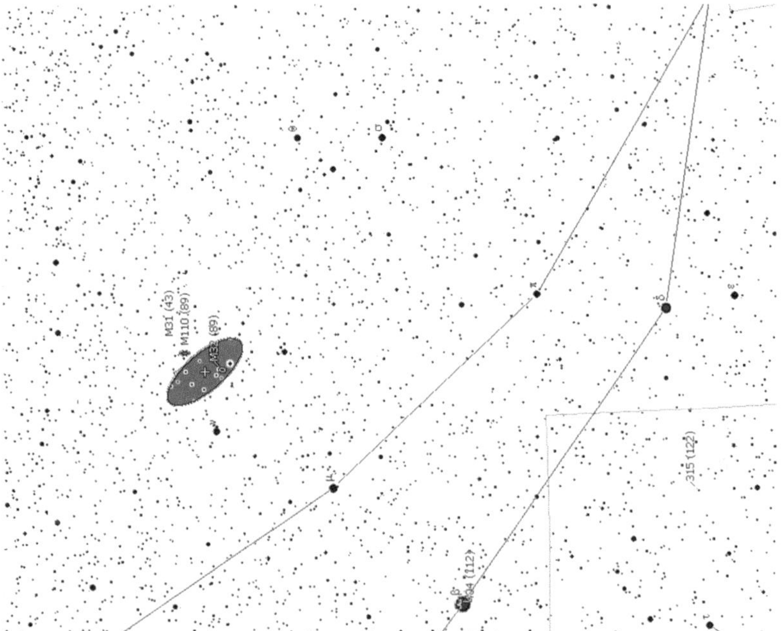

Fig. 3.1 Area near the Andromeda Galaxy. (Star chart courtesy of *Sky Tools 3* at www.skyhound.com)

There are other effective alternatives to a magnifying finderscope. Many choose to use a red dot or similar illuminated reflex finder. These are non-magnifying, offer a wide field of view that matches what you see with the naked eye, except it adds a red dot or similar point of reference with which to center your target. Most include an adjustable brightness level so that the targeting feature does not overshadow the stars in the field of view. Some of these, although good targeting devices, are too short when mounted on the telescope tube, making it difficult to get behind it comfortably enough in order to see through the window properly. If you do choose this kind of finder, look for one that sits higher up off the telescope tube, providing a much more comfortable viewing position. A popular version offered by many telescope accessory stores now is the illuminated finder that has a rectangular window with rounded edges. It offers four different targeting options—a red dot, a circle, a crosshairs, and a combination crosshairs and circle.

Another standard piece of equipment that has been in use by amateur astronomers for many years is the Telrad. This is a little larger than other finders and works especially well on medium- to large-sized 'scopes. Its design is a bulls-eye, the inner ring is a ½° in diameter, the middle ring is 2°, and the outer ring is 4° (excellent for mimicking various eyepiece field of views). For people who want a more compact design, or who own Schmidt-Cassegrains or refractors, the QuikFinder may be more appropriate, as it sits much taller although the outer ring is only 2° in diameter. Both of these devices offer an adjustable brightness display and provide an upright, correct image view to match the naked eye. It can be used as your sole targeting device, although some use it along with a regular magnifying finderscope. One additional consideration if you observe primarily in an area with moderate to heavy light pollution, where you can only see down to magnitude 4 or so. You may find it better to use a magnifying finder that will make more stars visible to your eyes for targeting. Otherwise, a reflex sight with no magnification will only show you the light-polluted sky as you see it with the naked eye, so you will be hard pressed to see very many star patterns to help locate objects.

Star Charts

For today's amateur astronomer, there exist a wide range of tools to reference the positions of the stars. In Chap. 1, we looked at several software options, including both smart phone apps and also desktop computer software. Although it seems our world is headed toward a digital majority, there is still a lot of value in holding a reference in your hands that can be marked and otherwise manipulated, free from the need for electric power. Let's take a closer look at some of the available star atlases you may want to use.

One of the most enduring atlases is *Norton's Star Atlas*, originally authored by Arthur Norton. It first appeared in 1910 before the boundaries of the constellations were officially set by the International Astronomical Union in 1930. Since then it has been revised many times; the latest is the 20th edition (edited by Ian Ridpath) released in 2004. It is an excellent reference work for people new to astronomy, covering subjects such as choosing a telescope, observing the Moon and planets, as well as several astrophysical topics. The atlas plots stars down to magnitude 6.49, including double stars, variable stars, and more than 600 deep sky objects. Each chart is accompanied by a table containing a list of interesting objects plotted within the chart, and brief descriptive notes are also listed beside most of these objects. The only drawback might be in regards to the charts. People who are new to astronomy may be reluctant to use them for field use, due to the lack of lines connecting the familiar shapes of the constellations.

The *Cambridge Star Atlas* by Wil Tirion is another good choice. The maps are easy to read as its dimensions are slightly larger than an 8.5 × 11 in. sheet of paper. The 4th edition includes a section devoted to the Messier catalog, featuring eight pages of charts wherein every Messier object is plotted. After this, charts covering the constellations are next, with lines connecting the asterisms and constellation lines. The book includes stars plotted down to magnitude 6.5. This atlas also contains tables featuring a sampling of stars, nebulae, and galaxies. There are also several all-sky maps. Overall this is an excellent tool to help you learn the constellations and some of the interesting objects they contain. It is very portable, spiral bound, and lightweight. *Sky & Telescope's Pocket Sky Atlas* is similar but has smaller dimensions and contains 30,000 stars plotted down to magnitude 7.6.

The Constellation Observing Atlas (by Grant Privett and Kevin Jones, Springer), is well designed and practical. Each section includes a short history of the featured constellation, a list of noteworthy deep sky objects, and a color map of the constellation along with those that border it. Stars are plotted down to magnitude 8.0, and charts include both the northern and southern hemispheres. *Guidebook to the Constellations* (by Phil Simpson, Springer) is also well suited for learning the constellations. Its approach is to use storytelling to take you on a journey through the stars. At over 800 pages it's much more hefty than most constellation guides, but the additional information is well worth having in one reference. Each section includes an interesting saga of the featured constellation, a list of prominent stars and noteworthy deep sky objects, and a map of the constellation along with bordering constellations. The book makes identifying constellations easy by providing charts that use lines to connect the shapes that form each character. It is well written and beautifully illustrated.

You will definitely want to take a look at the highly recommended *Sky Atlas 2000.0* (by Wil Tirion and Roger W. Sinnott, Cambridge U. Press) at the next star party you attend. Whether you use a 4-in. (102 mm) 'scope or even a 10-in. (254 mm) 'scope you'll find that it contains a wealth of stars and objects for you to locate and view, going beyond the beginning stages of deep sky observing. It contains 81,312 stars as faint as magnitude 8.5, and displays approximately 2,700 deep sky objects. About 200 stars are labeled with their common names, others are referenced by Bayer designation, Flamsteed number, or both.

For deep sky objects, galaxies as faint as magnitude 13 are plotted, and globular clusters down to magnitude 11. *Sky Atlas 2000.0* is laminated to protect it from dew and so is very practical for use in the field. Its size is a generous 19 × 14 in. (48 × 35 mm), making it much easier to see than the smaller atlases. Each page is clearly labeled with a key of how the objects are shown on the charts, and each page also labels right ascension, with declination along the sides. It comes in several versions: black stars on a white background, white stars on a black background, or full color. For a little less money you can purchase a non-laminated atlas in either of the black and white versions.

For amateurs with large telescopes of 10 in. or more (254 mm), the *Uranometria 2000.0 Deep Sky Atlas* (All Sky Edition 2012 by Tirion, Rappaport, and Remaklus, Willmann-Bell), is considered one of the best desk references for deep sky observing. The entire sky (both northern and southern) is covered in just one volume. Its charts are based on the latest data from the Hipparcos and Tycho catalogs, contains over 280,000 stars down to magnitude 9.75, and plots in excess of 30,000 deep sky objects. Its charts are extremely easy to read and user friendly, as each page is clearly marked with the range of coordinates and constellations that are covered. Its companion, *Uranometria 2000.0 Deep Sky Field Guide*, lists data for all of the deep sky objects plotted on the charts. Any object can be looked up by its catalog number or common name. Some areas of the sky (such as galaxy clusters) that normally would be too congested to display clearly have been enlarged and enhanced to display stars down to magnitude 11.5. Both the atlas and its companion deep sky field guide, are lightweight given their size, and are beautiful books printed on high-quality paper. Since it is not possible to laminate an atlas of this immense size, it may be subject to the dew if left out in the open at night for a long period of time. All things considered, this volume is a must have for deep sky enthusiasts.

Many other atlases are available, but before purchase it's a good idea to look it over at your local library, or ask to take a look at the ones in use by amateurs at star parties. Some like to have more than one atlas in their collection, depending on what their requirements are for a particular observing night. If you are going to have a more casual session with just binoculars or perhaps a small telescope, an atlas covering stars to magnitude 6 or 7 night is small and easy to grab and go. If you are going to seek a lot of deep sky targets, a more comprehensive reference may be necessary for the evening.

Eyepieces

Your telescope is like a window to the universe, and for it to show you its best, eyepieces must be given important consideration. A good set of eyepieces is an essential part of every observer's astronomical equipment. Using the right ones can help you to develop a new appreciation for your view of celestial objects.

There have been many different types of eyepieces designed over the years, including Huygens, Kellner, Orthoscopic, and Plössls. To judge the quality of an eyepiece, first be sure that your telescope has been set up properly, including accurate collimation. Then you can make a reasonable assessment of the images you see in your eyepieces. There are four basic problem areas in eyepieces to avoid:

1. Field curvature can be described as an image that cannot come to a focus in both the center and the edge of the view at the same time.
2. Coma in an eyepiece will make stars near the edge of the field of view look like comets. This kind of aberration is common in fast telescopes with a low focal ratio of f/5 or lower. However it should not be noticeable in 'scopes that have a high focal ratio. If coma is still present, it may be because of the eyepiece.
3. Distortion in an eyepiece occurs when the magnification appears to be higher nearer the edge as compared to the center of view.
4. Chromatic aberration causes stars to appear with an unnatural color distortion.

For the most part, modern eyepieces should be free of most if not all of the above mentioned problems. Still, it is possible that some have acquired an eyepiece of inferior quality. For example, eyepiece designs such as Huygens are of a much lower quality and should be avoided in favor of a good Plössl or other multi-element wider angle eyepiece. In the last 30 years or so the use of multi-element eyepieces has become popular because of the higher quality and wider field of view. The Meade line of Plössls are very good and have a decent 52° field of view.

If you are using a medium-size 'scope, more than likely it came with one or two decent eyepieces, perhaps a couple of Plössls. These are fine to get started with, and as you gain experience with observing you will no doubt want to expand your collection. Here are a few areas to explore.

First, you will want your set to include a wide range of useful magnifications. Two medium-powered eyepieces are good, but you will want a high-power eyepiece to reveal more detail for planetary viewing and one lower power eyepiece with a wide field of view. Please note, however, that higher magnification does not always mean you will see more detail. When increasing the magnification, the view in the eyepiece grows dimmer, making it more difficult to appreciate deep sky objects, most of which are already dim. Sometimes it is better to use a lower power eyepiece, as long as it still allows you to see the details in an object. You will be able to enjoy a wider field, with images that are brighter.

A truly wide field eyepiece will provide some of the most pleasing views of star fields, as well as deep sky objects such as nebulae and star clusters. Recall the best views of star fields that you have had through binoculars, which in most cases is anywhere between a 5° and 8° field of view, and compare that to an eyepiece that may only provide a true field of view of about ½° to 1°. You may be disappointed, hoping for broad views of deep sky objects, but only seeing a much more narrow view. Having a set of eyepieces with none that show a nice wide field of view is like shutting one of the panes on your window to the universe. Now imagine being able to at least double that field of view to about 2°. This makes a dramatic improvement on what you see in your observing sessions.

Let's take a look at what *exit pupil* is, so as to understand more about what to look for in an eyepiece. The exit pupil is the size of the image formed by the eyepiece and is directly related to the focal length of the eyepiece. The maximum practical focal length of an eyepiece for your telescope can be determined by a general guideline, that it should be no larger than 7 times the focal ratio of your 'scope, although opinions on this vary. This is because the maximum pupil size for most people is 7 mm, and the figure shrinks with age. So if you have a 6-in. (152 mm) f/8 refractor, use an eyepiece with a focal length of no more than 56 mm. Otherwise the size of the light beam from the eyepiece (the exit pupil) that enters your eye will exceed the size of your pupil, and much of the light (essentially from your telescope's aperture) would be wasted. We can also directly calculate the exit pupil by simply dividing the focal length of the eye-piece by the focal length of the telescope.

In a reflector you can see the bad effects of using an eyepiece with a focal length that is too large—you see a black spot in your field of view, which removes some of the visible light from the 'scope's aperture and is also very annoying. For example with an f/5 reflector telescope, a 41 mm eyepiece would yield an exit pupil of 8.2 mm, and this large exit pupil would lead to a visible black spot caused by the secondary mirror obstruction. Refractors do not have this problem, since there is no central obstruction, so it may be possible to exceed an exit pupil of 7 mm if you have need for an especially wide field of view.

Another important factor to consider is *eye relief*, especially if you must wear glasses when looking through the eyepiece. Eye relief is the furthest distance your eye can be from the eyepiece and still comfortably see the image. Most modern wide field eyepieces provide a generous amount of eye relief of 12–18 mm or more. If this is a concern for you, always check this specification before buying an eye-piece. You might also try a using a Barlow lens for higher magnification, since it will give you more eye relief than using an eyepiece of a shorter focal length that provides the same magnification. Two great lines providing very long eye relief of 20 mm are the Celestron Ultima Duo eyepieces, and the TeleVue Delos eyepieces.

You'll recall from Chap. 1 that true field of view can be calculated by dividing the apparent field of view by the magnification of the eyepiece. In the above example a 25 mm Plössl with an apparent field of view of 52° will give a magnification of 49× and a 1° true field of view in the f/8 refractor. (152 mm aperture × f/8 = 1,216 mm focal length of the telescope; 1,216 mm ÷ 25 mm eyepiece = 49×; 52° apparent FOV ÷ 49× = 1°). Although this is not bad, by spending more on this important fea-ture, the field of view can be increased to 2.3° with a 41 mm eyepiece with an apparent 68° field of view. What does it cost? Considerably more, but well worth it when you see the view!

Long focal length eyepieces with large apparent fields of view do cost more. Some of the best available are made by TeleVue, well known in the amateur astronomy community for producing some of the finest eyepieces available. The Ethos line of eyepieces made by TeleVue offers an incredible 100° field of view, and come in focal lengths ranging from 21 mm down to 6.0 mm, and the 4.7 mm and 3.7 mm both have a 110° field of view. Other high quality options are also available from other companies.

Fig. 3.2 An assortment of eyepieces that are commonly available. (Photo by the author)

As an alternative consider the line of super wide field eyepieces made by Orion that have an apparent 70° field of view. For around $100 (£65) its 38 mm super wide angle provides a 2.1° true field of view in the above example of the f/8 refractor. Eyepieces made by Explore Scientific (such as the 14 mm 100° apparent field of view eyepiece) are of high quality and also provide excellent wide field views, even when compared side by side with more expensive ones. Meade also makes very good Plössls, wide angle and super wide angle eyepieces. Keep in mind that most super wide field eyepieces usually have 2-in. barrels. If your telescope does not have a 2-in. focuser you may need to change it out to be able to enjoy most 2-in. barrel eyepieces. All things being equal, it will generally come down to cost and the most obvious discernible differences in quality. As is the case with other equipment, ask to take a peek through the eyepiece of others at a star party. This will help you narrow down your preferences after seeing the view for yourself (Fig. 3.2).

Dew Prevention

During long periods of observing you will eventually encounter images that have become obscure. The problem is dew. When there is a high level of moisture in the night air, and the temperature of your telescope falls below the dew point, drops of condensation can form on your optics and any outside surfaces. This will end your observing session unless you have some kind of dew shield. Telescopes most susceptible are refractors, Maksutovs, and Schmidt-Cassegrains. The lenses must be protected to prevent condensation from forming. The primary mirror of a Newtonian reflector usually remains dew-free because it sits so far down the tube that its temperature does not fall below the dew point. However, the secondary mirror is at risk along with the finderscopes of all telescope models.

Even if your telescope lenses or mirrors remain dew free, dew on your finderscope means trouble because you will not be able to sight on a target from a wide field of view. Some 'scopes have small extensions beyond the lenses that do help, but often more is needed for longer observing periods. Dew shields can be purchased at a reasonable cost. For those who prefer making their own solutions with materials, a trip to the art supply or hardware store may yield some acceptable options. Look for flexible foam or matte material that can be shaped into the size of your telescope's diameter (with a few millimeters to spare) and then attached to the outside of the tube. One such material is ABS plastic. It is thin, durable, and sold in sheets at a very reasonable cost. With it you can shape your own dew shield that should be adequate enough to prevent condensation from becoming a problem in most situations.

Other Helpful Tools

There are several other tools that will prove to be very useful to you on observing sessions. If you are using a star chart or atlas, as explained in Chap. 1, it is helpful to make a small plastic overlay with circles drawn indicating the field of view for your finderscope and eyepieces. This will limit any confusion as to just what you should be able to see in each instrument. It will also become extremely valuable for star hopping to your intended target.

Of high importance is being able to see in the dark, but this should never be done with a regular light. If you have been to a star party, you will remember how upsetting it is for someone's observations to be interrupted by car lights or a bright white light. At many star parties there is a dead zone beyond which such lights are strictly prohibited. As you may already know, the solution is a red light, which tends to cause the least amount of disruption to the eye's dark adaptation. That does not mean you should go out and buy a large bright red light. Look for a small handheld red light that offers a variable adjustment. With a variable adjustment you will be able to use just the amount of red light you need to see your charts or equipment and still preserve most of your night vision.

In the first chapter, the use of a laser collimator was explained. The laser collimator does not cost much and is a real time saver. Be sure to buy a new one, because it is possible to get a used one that is defective. For example, if the laser suffers a hard fall or similar damage, it can be rendered useless. If you collimate before leaving home, there will probably be very little change once you arrive at an observing site. A double check once you arrive and get your 'scope set up is always advisable, and using the laser makes that double check fast so you can quickly move on to observing.

As we discussed earlier, you should always bring a planisphere, smart phone app, monthly star chart or similar printed chart with you, according to your preference. This gives you a broad view of the entire night sky, and will help you to pinpoint which constellations are visible. If you are looking for something similar to a planisphere but a little more detailed, try the DeepMap 600 by Orion. It is a fold-out map of the entire sky, including all constellations, and has a separate area that covers the circumpolar stars. Additionally on the reverse side it lists the details of 600 deep sky objects, along with several interesting double stars and variable stars. This practical and durable map is made of plastic that is tear proof and weatherproof.

Most people are not in a position to acquire everything needed at one time. Instead, make a list of what you feel are the most important items to acquire in the least amount of time. A separate list can be made of those items that will take a little longer to accumulate. Gradually work your way down both lists until all of your needs have been met. By attending astronomy club meetings and checking online, you may also find great deals from people trying to sell equipment they have outgrown or no longer need.

Items of a Non-urgent Nature

Filters

In one way or another, a filter used on an eyepiece can help to improve contrast by removing something from your view. A *Moon filter* (also called a *neutral density filter*) helps to remove the glare you see when viewing the Moon. This will allow you to see a lot more detail than you would while struggling with the brightness that comes through when viewing the Moon through a telescope. The percentage of light that a filter allows to come through is stated in its specifications and can vary usually from 10 % to 50 %.

There are also *color filters* that come in a variety of colors used for planetary viewing. Much like a neutral density filter, while blocking some of the light, they add contrast by revealing details that respond to a particular color. Other widely used filters are those that reduce light pollution and those that enhance views of nebulae. This is accomplished by restricting some wavelengths of light from passing through the eyepiece, while at the same time allowing certain other ones to pass through. *Broadband filters* remove the wavelengths of light that come from urban lights (mercury-vapor and sodium lights), creating an effect of darkening the background sky. It is a general use filter for moderately light-polluted skies, sometimes referred to as a light pollution reduction filter. Although it is used by most as a general purpose filter for viewing all types of deep sky objects, it will dim your view of star clusters and galaxies. A better option would probably be to seek darker skies.

Narrowband filters go a step further, more aggressively removing light pollution but at the same time removing the glow of stars and galaxies. This allows for an excellent passing through of the light from diffuse and planetary nebulae, which is unaffected by these types of filters, since the narrow bands they transmit (ionized oxygen and hydrogen) are passed through to the eyepiece. The sky background will be much darker, and the resulting contrast will cause many nebulae to become more easily visible. Going even further there are *line filters* such as, *O-III filters,* and *H-beta filters*. Line filters only allow one kind of emission line to pass through from a particular nebula. The O-III filter passes only the ionized oxygen emission line, whereas the *H-beta filter* passes only the hydrogen-beta emission line. Most amateurs report that the O-III filter is great for enhancing views of the majority of the most elusive nebulae. On the other hand the number of nebulae enhanced by the H-beta filter is comparatively far less than those enhanced by an O-III filter, among them the Horsehead and California nebulae.

So although color filters can be extremely helpful depending on the situation, for the observer just beginning to move beyond first light, there is not an urgent need to obtain them, unless you are only interested in observing the planets. Quite a bit of detail can be observed in nebulae without any enhancing filters, and much more can be gained by exploring as much detail as possible without them before adding the whole set to your collection. As for the effect that the broadband, narrowband, and line filters will have on certain objects, the perception of enhancement varies from person to person. Everyone's eyes are different, and the only way to be sure of how you will perceive the enhancement a filter can bring is to test it for yourself.

GOTO Telescopes, Motor Drives, Digital Setting Circles, and Tracking Platforms

For people just beginning to become interested in visual astronomy, there is an immediate appeal to being able to turn on a computerized telescope and have it magically turn its lenses or mirrors to something spectacular to observe. There are in fact many benefits to using such equipment, such as taking a group of people on an automated tour of the skies, locating objects in areas that are heavily obscured by light pollution, and allowing photos to be automatically taken while you sit in the comfort of your home to name a few. Even so, it is by no means a requirement in order for you to enjoy all that amateur astronomy has to offer. In fact, you may do much better to start out in visual astronomy without this equipment, and earn the knowledge and skills with your own hands and eyes. This will bring immense satisfaction and a deeper meaning to what you are seeking to observe.

The desire for visual observing is similar to the desire that some have to prepare their own food, while eating out at your favorite restaurant can be compared to using a computerized telescope. You will be limited to whatever the computer shows you, much like ordering a dish at a restaurant—what you see is what you will get. On the other hand some enjoy preparing their own meals because this gives them the flexibility to include whatever they wish in the meal, a sense of accomplishment, and control over the ingredients. When you locate objects on your own, you maintain control of what you see. You will also know for certain that what you are viewing is the correct object because of the effort and extensive verification it took to get there. This also gives you the freedom to view other stars and interesting objects in the vicinity of your intended target. You will become familiar with the sky and come to understand what you see. The thrill comes from learning where the stars, nebulae, and galaxies are located and how to find them. Then finally, you arrive at your destination and view the object with your own eyes.

Many telescopes are mounted on an equatorial mount and may also come equipped with a motor drive for one or both axes. This allows the user to view an object while the telescope's motor drives automatically keep it in the field of view. In order to locate objects, you only need to turn the telescope to the coordinates found in your star atlas, using the 'scope's manual setting circles. Digital setting circles work similarly, and you may or may not have drive motors attached to the telescope. Both the GOTO features and equatorial mounts with drives can be found anywhere telescopes are sold. Digital setting circles are often used on 'scopes that are on an altazimuth mount, such as Dobsonian reflectors, making it easy to find deep sky objects as well as other objects of a more periodic nature, such as comets. At considerable additional cost, Dobsonian reflectors can also be fitted with GOTO capability. If you prefer tracking only, your Dobsonian can sit on top of a tracking platform that allows it to keep any object centered in the eyepiece for a long period of time.

Although having some or all of these features can be very useful and desirable, they are by no means a necessity for you to be able to enjoy astronomy. You can have many memorable observing sessions with the equipment that you currently own. Throughout this book you will continue to learn how you can develop the skill to observe anything you wish (within reach of your telescope's aperture) by using your 'scope along with a good star atlas by your side. Of course, eventually you may upgrade to a GOTO system or other tracking system of some type, especially if you plan to get into astrophotography. If you do, your appreciation for what you see will be greatly enhanced by the experience that you accumulate from locating celestial targets on your own.

Aperture Upgrades

Many of us have experienced aperture fever even if only temporarily. Someone comes to a star party with a 20-in. (50 cm) telescope, and you cannot help but wonder if it's time to upgrade your telescope aperture a little bit. For others, it can become a constant quest for a bigger 'scope. Although more aperture is desirable to see the faintest of objects, we do well to remember that the best 'scope is one that you will use the most often. It is not uncommon to see used large telescopes for sale in good condition that have only been owned for a short period of time (under 2 years or so). The reason is usually that the owner bought too much telescope, too soon. This can mean that the 'scope was too large and heavy. It can also mean that the person bought equipment that was too complicated for him or her to use. This reiterates the importance of learning to master the important basic equipment *before* gradually adding more advanced pieces to your array of astronomical devices.

If you do not have a clear view of the sky at your place of residence, a 'scope that is too large may present a real challenge to transport each time you want to observe, especially if your vehicle is not large enough to accommodate the size of the 'scope. A 'scope that takes two people to unload and set up can become not only challenging physically but it also may be a challenge to find an extra set of hands each time. When you are serious about an upgrade, consider one of the many designs for compact telescopes of very large aperture that have addressed this problem. However, this comes at a considerable increase in cost, and there are still other pieces of equipment that are necessary to acquire.

If you have a smaller telescope by comparison, perhaps something in the 4- to 8-in. range (102–203 mm), do not despair! There is plenty to see that is well within grasp of your telescope. Spend some time becoming adept at visual observing with what you have, and then after gaining valuable experience, you'll know when the time is right for an upgrade that you can truly manage and enjoy.

Dew Heaters, Cooling Fans, and Other Items

Eventually you may want something much more effective for dew removal. Some have used a little heat from a portable hair dryer or something similar, but this requires constant re-application, since the dew will keep forming. The solution is an electronic dew removal device. Many suppliers of telescope accessories sell these dew heaters. They require power (12 V DC), and have attachments that connect to each of your affected pieces of equipment, like an eyepiece or finderscope. Heated dew shields are also available. A little electrical warmth from this device can make your dew problems disappear for your entire observing session. If you will be observing for long periods of time, a dew heater is a must.

Large primary mirrors can retain heat, keeping them at temperatures above the outside air and causing tube currents. The best way to prevent this is to be sure that your 'scope is outside for some time cooling off so that it reaches ambient air temperature. These tube currents can degrade your images and mimic poor seeing conditions. A small cooling fan attached to the rear mirror cell or rocker base of a Dobsonian mount can help alleviate this condition. Fans can be purchased that run on either AC or DC power. You can also easily use a small computer fan and hook it up to a DC power supply.

When observing an object under high power, grabbing the telescope's focuser for adjustment can cause vibrations that disrupt an otherwise steady view. An electric focuser puts an electronic box between you and the focuser so the 'scope is not touched, keeping the image steady and removing the possibility of completely moving the 'scope's position altogether. The electronic focuser allows for incremental focus adjustment, but of course at a much higher cost. This is certainly not an essential accessory, but if you plan to do a lot of planetary viewing under high magnification, you may want to put this on your wish list.

Having a power supply is not a requirement unless you wish to use some of the features we've covered in this section or unless your telescope is designed to require it. If you are skilled with wiring and electricity, you can rig your own power supply. Otherwise rechargeable DC power supply options are now commonly available. Celestron and Orion offer several great choices. The power supply is often combined with an emergency light, radio, and jumper cables—which are very nice to have in case your car needs a jump after a long drive to a dark sky site. Review the power supply's specifications to be sure you do not exceed the amount of amps that it is designed to support.

As optical quality has increased over the years, many amateurs have grown to prefer the views that a *binoviewer* can provide. Observing with both eyes adds a more multi-dimensional effect and is more comfortable because you no longer have to keep one eye closed while looking into an eyepiece. TeleVue, Antares, and several others offer binoviewers. Denkmeier's Binotron-27 Binoviewer has been described by some as one of the best on the market. It comes with a matching set of eyepieces, allows individual eyepiece focusing, and offers three different magnifications with the same pair of eyepieces.

Useful Household Items

Since we are normally so focused on bringing our telescope and eyepieces to a viewing site, it can be easy to forget the little things that are also necessary. Here is a list of some of those necessary items that you more than likely already own:

- A CARRYING CASE FOR YOUR EYEPIECES. You will need something to secure your eyepieces and protect them from the elements. A small suitcase or similar size bag or briefcase can be converted for this use. Cotton or other soft materials can be used to provide a soft cushion for the eyepiece boxes that you place inside. If you prefer, sellers of telescope accessories offer very attractive choices of durable eyepiece cases that are lined with foam to protect your equipment.
- An OBSERVING CHAIR. Like the eyepiece carrying cases, there are chairs made specifically for astronomy. They have an adjustable height allowing you to remain seated while viewing more areas of the sky. Any lawn or outdoor chair will do almost as well. Camping chairs are collapsible, so they fit easily in your car and are very inexpensive. Remaining in a seated position for as much of your observing session as possible is more comfortable and allows you to see more detail since you will be more relaxed.

- ALLEN WRENCH, FLATHEAD AND PHILLIPS SCREWDRIVERS, and any other tools you may need specific to the type of telescope you own.
- SPARE BATTERIES for any equipment you use.
- If you use a reflector a LARGE SHEET OR TARP is useful to spread down onto the ground before setting up your 'scope. This will help keep your optics clean by limiting the amount of dust that is stirred up.
- A piece of PLYWOOD found at any hardware store helps to create a more level surface on which to set up your telescope.
- If you like to spread out and have some space for charts, eyepieces, and so forth, you should bring along a SMALL FOLDING TABLE. A small red light lamp can be placed there that is aimed downwards, and the table can also become a great portable work station, especially if you try your hand at sketching what you see in the eyepiece.
- A small NOTEBOOK AND PENCIL for recording your observations or sketching.
- A SMALL WAGON can be useful by serving a dual purpose. First you can use it to haul equipment from your car. Then, once your 'scope is set up, it can also serve as a table for your accessories.
- There's no need for hunger to cut your observing session short. Bring a few healthy SNACKS AND LIQUIDS to help maintain your energy level during a long night of observing. However, alcoholic beverages and large amounts of caffeine are not recommended because they may dehydrate you and limit your night vision from performing at its highest potential.
- Dobsonian owners may want to bring a little LUBRICANT (such as silicone) for the bearings of the mount. This will ensure smooth altazimuth motions on your scope.
- MOSQUITO REPELLENT is an absolute must.
- EXTRA CLOTHING. Temperatures at night tend to fall below what most people expect. It is always wise to bring an extra layer of clothing just in case.
- If you will be driving far to your observing site, be sure to fill up your GAS tank and always remember to take a fully charged CELL PHONE. Also, be sure to let someone know where you are going, even if you will be meeting others at an observing site. JUMPER CABLES are also a good idea, as the last thing you would want is to be stranded after a long night of observing.

At first glance it may seem like a long list, but once you gather these items a few times, you will find that your loading time may be cut down to 15 min or less. Everyone's list is different, and you may not use everything listed here, and you might also end up adding a few things that you need. If at all possible, on the day you will be observing or at least that same week, practice setting up your telescope. If you wait until dark, whether you observe at your home or somewhere else, it can be frustrating to try and work with unfamiliar equipment and tools. Save yourself time and energy by a practice setup or two, which can also reveal any potential issues with your equipment. Then these issues can be sorted out before you have used most of your observing time to address it.

Skills, Attitudes, and Habits

Consistency

Making a success out of each of your observing sessions will be very difficult if you are seldom out under the night sky. Sporadic observing is just not conducive to developing a sharp memory of the constellations or the other important skills you need to develop. It is much better to develop a habit of regular observing, even if each session is short. Perhaps over time, you can extend the length of your sessions so that you can observe in an unrushed manner, taking your time to uncover those subtle details that elude the inexperienced.

Everyone has a different schedule and physical tolerance for being outside at night, but the best way to be certain of your progress and success is to be out as often as possible. In the very least, one night of viewing through the telescope and another with binoculars each month should suffice to get you started off in the right direction. Most astronomy clubs are geared toward this kind of schedule, as they often have club meetings once a month and at least one day a month near the new Moon when everyone gets together to observe. Before you know it, your time under the stars will grow in frequency and length. You will end up having more fun than you thought possible, and stretch far beyond any pre-conceived limits to your capabilities or range of involvement in astronomical activities.

Star Hopping

Clear skies are a precious commodity. Both professional and amateur astronomers treasure each night they are able to have a clear view of the cosmos. At star parties and outreach events, you will find several people who seem to float effortlessly from one beautiful object to the next. These amateur astronomers are star hopping, moving from one familiar star or group of stars to the next. Some are so good that they can land a target with just one smooth motion. Perhaps you have wondered how in the world someone can see something so tiny in such a vast sea of light.

At some point in the past we may have looked at the stars and seen only bright dots that appeared to be strewn about at random. Over time we began to see shapes that we later came to realize make up the constellations. With practice, similar structure can be perceived as you learn to find your way not only in the sky as a whole but also within a much smaller field of view, such as in your finderscope or eyepiece. Just as we have the ability to recognize faces, we have and can develop *star pattern recognition*. This skill will help you to find your way around the sky more quickly and efficiently.

There are billions of stars in our galaxy, so the term star hopping can seem a little daunting, maybe somewhat inefficient. Although it is true that at times we may move from one star to another, more often it is actually several stars that we see and recognize in relation to each other. That is what helps us to advance from one position to the next until we are in close proximity to our target. Then, we find smaller star patterns, in effect miniature constellations, and we no longer feel inundated by so many individual stars.

We do not need to dread star hopping if we truly understand it. It is not so much star hopping but rather a matching of star patterns made up of several stars. It is a step by step process that begins first with star pattern recognition, and continues by star pattern jumping to the intended destination. Once you have located the object, you can take a stellar excursion around the target and find even more celestial treasures. From this point we can comprehend that more than one path can be selected to find an object and therefore after learning one passageway to an object, you can teach yourself any one of several others.

Soon we will review several examples of this concept beginning with some of the more easy targets, and then gradually move on to more challenging ones. To prepare yourself for these celestial tours, begin looking over groups of stars on a star chart. Try to see the stars in groups of three to no more than ten stars. What shapes does your mind's eye see in them? Are they animals, places, man-made objects, or people? Maybe they are simply geometric shapes like polygons (regular or irregular), squares, triangles, and lines. After a few minutes, leave that part of the sky chart and review a different chart. Try to see shapes and patterns on the new page. Now go back and review the first one. Can you remember the patterns that your mind created? Even if you only remember one or two, you have just taken the first step to developing star pattern recognition beyond the shapes of the major constellations. With continued practice you will be able locate some very hard to find objects, and will be greatly rewarded for your diligence.

Patience

Nothing will take you further in astronomy than patience. It is needed to grasp important concepts. Finding small dim objects in places all over the celestial sphere also requires a lot of patience. You need to learn to give yourself credit for each bit of progress that you make. Great things seldom happen overnight, but if you are patient they seem to happen that way. Becoming a visually wealthy observer is attained by the accumulation of small accomplishments over time. As you move forward in this book and review tours of deep sky objects there may be times when you just do not find the target at first. This is normal for all observers. Stop, take a deep breath, and retrace your steps. With a little patience you will eventually be rewarded with finding the object. With each observing experience your patience will develop into a habit that you will apply without giving it a second thought. Then it will be in those quiet, patient moments that you will see faint objects and details within them, that up until that point you had only read about. Those moments will be the unexpected times, the special times of discovery.

Record Your Observations

There is no telling just what stunning sights you may encounter, so it is advisable to take a notebook with you to record what you see. If you prefer, you can use a pre-formatted observing sheet or log such as is used by many amateur astronomers. A sample observation log is provided in Fig. 3.3. A common way to notate your observing conditions listed on the log for a particular session is to use a rating system of 1–5, with 1 meaning poor, and 5 indicating excellent.

Some astronomy clubs encourage members to explore a variety of deep sky objects through a group of observing programs. You can take a look at some of these programs on the Astronomy League's website (www.astroleague.org). In order to validate whether or not all of the objects on each list have actually been seen, they are required to be recorded on a log sheet that includes all of the pertinent information, such as seeing conditions, date, time, etc. When you write down what you have seen, much more detail is recalled. Better yet, a brief sketch of what you see will help your eyes to focus on the tiniest features.

Astronomers have been using this form of record keeping since the time of Galileo. Famous astronomer E. E. Barnard regularly drew detailed sketches of Jupiter with its many features, such as the Great Red Spot. Many amateurs astronomers who sketch share their work on websites such as Astronomy Sketch of the Day (www.asod.info). So instead of simply telling someone about the things you saw, you can show them. Many observers who sketch at the telescope post their work online for their friends or others to enjoy. The Association of Lunar and Planetary Observers encourages sketching and relies on the information that is gleaned from the drawings that its members submit. Many drawings submitted by amateur astronomers are printed in the Association's quarterly journal *The Strolling Astronomer*.

Observe Other Amateur Astronomers

One of the best ways to learn amateur astronomy is to watch others do it. What equipment do they use? What kind of objects are they observing? A star party sponsored by your local astronomy club is sure to have several people there who would be more than willing to take you under their wing for the evening. Like any gathering of people there is sure to be at least one personality that you will find to be a good match for you. Perhaps there is someone like the person described earlier, who moves easily through various objects in the night sky. That's definitely someone to watch for a while, picking up suggestions on everything astronomy related. Do not be afraid to ask questions. Just asking "How did you find that?" can lead to a very practical session that in the future will help you to remember how to locate one of your favorite objects (Fig. 3.4).

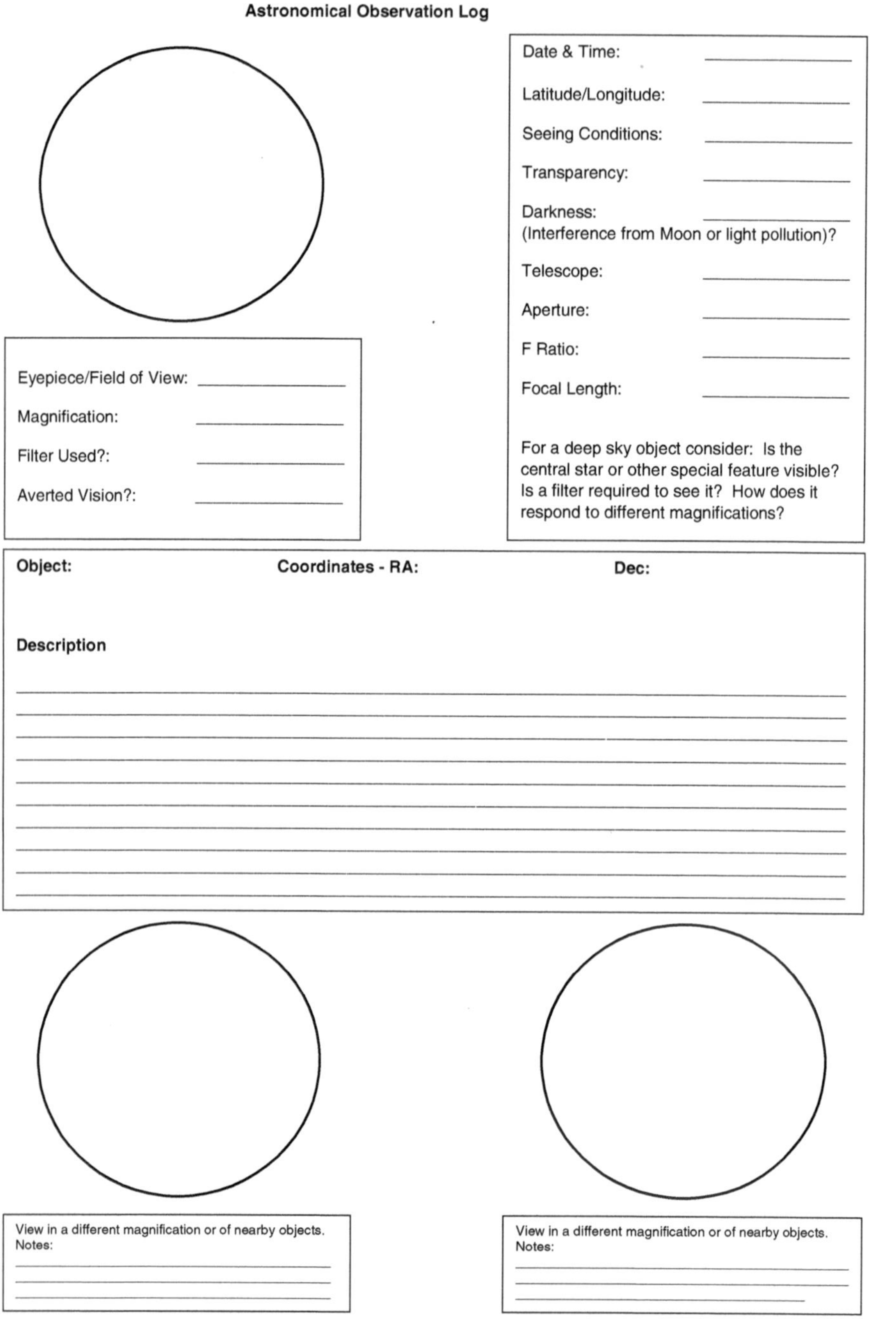

Astronomical Observation Log

Date & Time:	_____
Latitude/Longitude:	_____
Seeing Conditions:	_____
Transparency:	_____
Darkness:	_____
(Interference from Moon or light pollution)?	
Telescope:	_____
Aperture:	_____
F Ratio:	_____
Focal Length:	_____

Eyepiece/Field of View: _____

Magnification: _____

Filter Used?:   _____

Averted Vision?:   _____

For a deep sky object consider: Is the central star or other special feature visible? Is a filter required to see it? How does it respond to different magnifications?

Object: **Coordinates - RA:**   **Dec:**

Description

_____
_____
_____
_____
_____
_____
_____
_____
_____

View in a different magnification or of nearby objects.
Notes:
_____
_____
_____

View in a different magnification or of nearby objects.
Notes:
_____
_____
_____

Fig. 3.3 A log such as this one can be very useful for recording observations at the eyepiece or can be transcribed later from a voice recording. (Illustration by the author)

Fig. 3.4 One of the best ways to become better at amateur astronomy is to learn from others. (Photo courtesy of TWAN/Babak Tafreshi. Used with permission)

Find a Dedicated Mentor or Observing Partner

So now it's time to go beyond simply observing someone and seek out a mentor. This is someone that you can meet with on a regular basis or at least several times so as to build up your observing skills. Find someone whose teaching personality matches your learning style. That is why it's a good idea to observe experienced amateurs first, so you can get a sense of how they teach. One person may like to share a lot of information very quickly, maybe covering 20 objects in one night, whereas someone else may take a more leisurely stroll through the galactic neighborhood, sharing interesting details with you along the way. There is no right or wrong approach—there are simply different ways of approach. In fact you may prefer to have more than one mentor or observing partner. Some find it easier to absorb information that is expressed in various ways by different people.

In any activity, humans usually respond well when surrounded by mutual encouragement, and astronomy is no different. On a cold wintery night, or a hot summer evening full of mosquitoes, having company can help you maintain a sense of humor and keep the fun going when outside in less than perfect conditions. More importantly, a mentor or observing partner can help keep you motivated to keep going and not give up. They are there to answer questions and help you find the answers if they do not know them. The interest you show in learning astronomy will be encouraging to your mentor as well. Teaching you will help your mentor to hold onto their knowledge and is one of the best ways to improve their own prowess at the eyepiece.

These are just a few of the more essential areas that you should be aware of and focus on. If you maintain a consistent effort to develop these skills and habits, it will lead to many successful observing sessions. You will become a knowledgeable amateur astronomer who enjoys the night sky, instead of one who dreads it for lack of understanding. Your equipment will earn the distinction of becoming well-worn through regular use, as opposed to collecting dust in a closet only to be sold later as "used like new" to the next newcomer. Who knows, you, too, may one day pass on the wonders of the cosmos to the next generation.

Helpful Things to Consider

Why is it important to have an appropriately sized finderscope? What particular skills, attitudes, or habits do I need to develop the most? How can a high quality eyepiece or two affect the quality of my observing sessions?

Before considering an upgrade in aperture, have I made the most of the telescope I already own? Will I be able to manage the additional size and weight of the new 'scope?

Here are a few reminders and additional points to keep in mind:

- Test out equipment first before purchase. This can be done with your fellow astronomers at star parties. Ask questions and find out how a particular accessory has performed for that person. Read reviews of the item in astronomy magazines or in online astronomy forums.
- Just because you happen to come across an item that is favorably reviewed, it does not mean you must have it. Consider carefully whether the equipment is right for you.
- Prioritize your equipment purchases. Organize them into things you need more urgently, and then those other things that can wait to be added later. For example, before upgrading to that large scope you want, it may be better to add a high quality wide field eyepiece to your collection. It will greatly enhance your viewing pleasure and make it easier to locate objects.
- To successfully navigate through the constellations, develop your star pattern recognition. This same skill will help you when viewing stars in a very small field of view in the eyepiece. There you will be able to star pattern jump until you reach your destination.
- Maintaining an observing record is a good way to track your progress over time. Recording what you see will also help you develop your ability to see more detail in objects that you may have missed before. A sketch or two is also sure to create a much more memorable observing session.

By now it should be clear that it is important to have the right equipment for successful visual observing, and that it does not require spending inordinate amounts of money. In fact, a lot of what you need will not cost you a cent, but may simply take time to develop. On the other hand there may be some helpful tools that you will want to acquire over time. The right equipment for one person is not necessarily the best for another. You will have to consider what your own observing needs and desires are, according to the areas of amateur astronomy that you are most intent on exploring.

Now it is time to consider some interesting celestial sights that you will be able to find. Everything that has been discussed up until now will be very useful when you read the next three chapters. To help prepare you, we'll first cover plenty of helpful information about the use of averted vision and the value of preparing your observing sessions. Then we will take a tour of many star clusters and nebulae that you should find very intriguing.

Chapter 4

Beehives and Eskimos: Searching for Star Clusters and Nebulae

It is thrilling to contemplate that we observe many of the same stars that our ancestors did long ago, who named or described the objects they saw in the sky. For example, what we refer to today as the Double Cluster in Perseus (NGC 884 and NGC 869) was identified on celestial charts of the eighteenth century as being located in the sword hand of Perseus. The Greek astronomer and mathematician Hipparchus (190–120 B.C.) was probably the first to record seeing these naked-eye clusters. He referred to them as a "cloudy spot." Later, Ptolemy (A.D. 87–150) would call them a "dense mass."

Some of the names and descriptions from ancient times have endured until today, still evoking wonder and mystery as we seek to explore these objects with our modern scientific instruments. Other names have changed, but we still find that our references to them are often very apt descriptions of their appearance through a telescope. The Beehive Cluster (M44) is an open star cluster containing more than 1,000 stars, and is approximately 580 light years distant from Earth. Its brilliant stars are grouped together in a way that somewhat resembles a gathering of bees in a beehive. The Eskimo Nebula (NGC 2392) is a planetary nebula situated about 2,900 light years away in the constellation Gemini. When viewed in a powerful telescope it looks as if it is a face surrounded by a fur parka. As you begin to examine these and other objects for yourself, you will develop your own unique perspective of what they resemble (Fig. 4.1a, b).

This chapter will help you to explore two different kinds of deep sky objects: star clusters and nebulae. As a new observer, maybe you attended a star party and were amazed as you watched a veteran amateur astronomer find object after object with ease. Acquiring that kind of skill takes time and experience. However, you, too, will become effective at finding and observing deep sky objects, if you take

© Springer International Publishing Switzerland 2015
D. A. Jenkins, *First Light and Beyond*, The Patrick Moore Practical Astronomy Series,
DOI 10.1007/978-3-319-18851-5_4

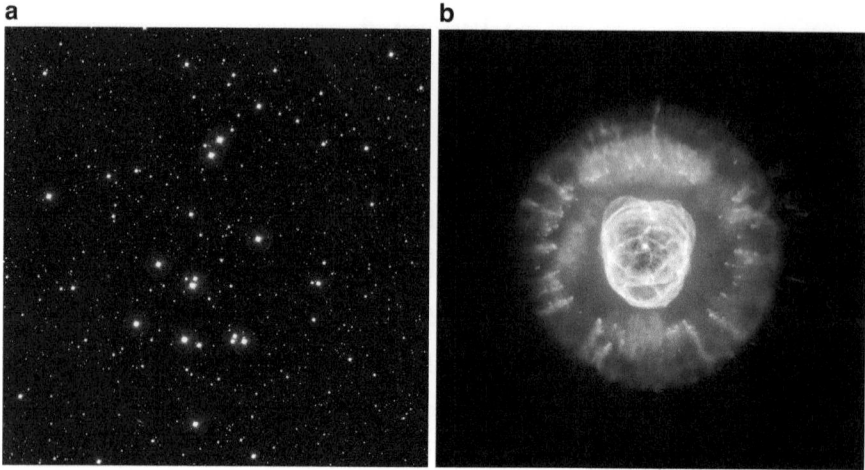

Fig. 4.1 (**a**) The Beehive Cluster (M44) in Cancer (Image courtesy of NOAO/AURA/NSF.) (**b**) The Eskimo Nebula (NGC 2392) in Gemini (Image courtesy of NASA, Andrew Fruchter and the ERO Team [Sylvia Baggett, STScI; Richard Hook, ST-ECF; Zoltan Levay, STScI])

time to understand and master certain essentials. Some of these essential skills were covered in the first few chapters, and now we will review some of them and expand our understanding in the context of deep sky observing. Next we will examine how to find and observe some of the most stunning star clusters. Then we will move on to finding nebulae and demonstrate how you can be successful in exploring them. For the most part, this chapter will focus on hints to help you locate and observe celestial phenomena. For now, just what are some of the essential skills and ideas?

Know the Sky

In order to successfully observe the night sky, you should first become comfortable with your surroundings. Learning at least the more prominent stars along with major constellations will prove to be invaluable. You may use digital setting circles or own a computerized telescope that features automatic slewing to the desired object, but even most of them require that you at first target two known bright stars into the computer's system as points of reference. There could also be a time when you are without power or observing with someone who does not have a computerized telescope, and in such a situation knowing your way around the sky by sight would be indispensable. It is also so much more rewarding to be able to discern what area of the sky you are looking at, and then find those targets that you may have missed before!

We have already discussed at length, in Chap. 1, the importance of learning the constellations. However, let's briefly recap three vital points and transition our way into a successful star pattern jump. First, begin naked-eye observing and learn the constellations with the aid of a planisphere, an astronomy app, or an all-sky star map. Using binoculars is the next step, as this is a good transition to the telescope. Binoculars will show you a much smaller field of view than your eyes, but at about 5° it is still impressive enough for you to marvel at beautiful star fields and swaths of nebulous clouds. The Andromeda Galaxy is also a spectacular sight in 8×50 mm binoculars or larger. More to our purpose here, though, binoculars will help you become comfortable with a field of view similar in size to your telescope's finder-scope, be it a red dot finder or a magnifying finder. Finally, once we switch to a telescopic view, our field of view will narrow considerably. Long focal length eyepieces will provide a much lower magnification, but the field of view will be larger. However, shorter focal length eyepieces yield higher magnification and consequently the field of view will be much smaller.

Before going to the eyepiece let's take some time to walk through a few constellations. If you are not yet very familiar with the sky, you may at first wish to choose an evening separate from the one when you will try using your telescope. This will let you focus on learning some of the easier stars to find, without the pressure of trying to find more challenging targets. Before stepping outside be sure that your clothing is appropriate for the temperature you will encounter, and take along a comfortable chair, binoculars, a red light, and your preferred format of an all-sky map. Once you are outside look up and begin taking in the view. Get comfortable, relax, and allow at least 20–30 min for your eyes to adapt to the dark. Over this period of time your eyes will develop an increased ability to see stars that are much fainter (more on this soon). At this point in your observing session, if you are not using an astronomy app, be sure the direction you are facing matches the way you are holding your printed map, planisphere, or all-sky map. Begin with the brighter stars you see, which are indicated on your map as those that are larger in size, while fainter ones are smaller. Most maps will contain a key listing the stellar magnitudes for your reference. You will no doubt recall from Chap. 1 that a star's magnitude is a simple way of indicating the apparent brightness of the star. The lower the number, the brighter the star is, and the larger it is drawn on the map. The larger the number, the fainter the star becomes, and it will appear smaller on the map.

Once you believe that you have found one of the bright stars indicated on your map, double check yourself to be sure. Compare the location of that star to its neighbors, and see if you can recognize the shape that it forms in relation to other stars nearby. As an example, let's take a look at a few stars in the summer sky that are visible in the northern hemisphere. Just after sunset, around late June through mid-July, high in the northeastern sky will appear the bright star Vega in the constellation Lyra. Vega is the fifth brightest star visible from Earth, and shines at magnitude 0.

See if you can find this star by comparing your view of the sky with your map or planisphere. From Vega, look 24° northeast and you will see a magnitude 1.2 star named Deneb in the constellation Cygnus. You may recall from our prior topic about approximating distances that this should be about the span of your hand fully stretched out, held at arm's length. From Deneb, look about 38° southeast and you will find a magnitude 1.0 star named Altair in Aquila. Vega, along with Deneb and Altair, form a recognizable pattern in what is called the Summer Triangle.

Now you can try to find and identify stars in other constellations. A good place to start would be with the seven main stars that form the Big Dipper. Then, while facing north, look between the Big Dipper and the Summer Triangle. Let your eyes follow a path south-southeast towards straight overhead, and you will find the constellation Hercules. Can you see the four stars that are referred to as the Keystone? To the west of Hercules you will see the bright −0.06 magnitude star named Arcturus in the constellation Boötes. This could be described as the tail of what appears to form the shape of a kite, along with the other brightest stars of the group.

Up until this point we have taken in our surroundings with just the naked eye, and have become more familiar with star patterns within several constellations. If you have not done so already, use your binoculars to make a closer examination of everything that you only perused with your eyes. How has your point of view changed?

For one thing your field of view is smaller through binoculars compared with using your eyes. However, the tradeoff is well worth it since you are gathering more light into your eyes and that means detecting a lot more stars. Some of the stars on your map that you were unable to make out now jump into view. Continue to view the rest of the stars as before, but note any differences in the patterns you see. Take time to reorient yourself to the asterisms in this intensified view, since many more stellar bodies now appear in the background. If you are observing from reasonably dark skies, not to be overlooked is the stunning view of what appears to be a long stream of starry clouds—the band of the Milky Way! You can see this beautiful, vast cloud of stars begin at the horizon in the northern sky, continue through the sky in a southern direction through the Summer Triangle, and go on through the constellations Serpens Cauda, Sagittarius, and beyond. The Milky Way band is an excellent region in which to spend time exploring with binoculars.

Plan Your Sessions

In our last chapter we discussed the importance of having the right equipment and making sure you learn how to use it *before* going out under the sky. So, too, when it comes to your observing session with your telescope, it would be a good idea to take a little time and plot out what you will be looking for. In addition, a well-planned observing session can lead to unexpected surprises that you might find along the way. This is simply because if you are looking for a specific list of targets that fit a particular description, you are more likely to notice if you stumble on an object that is different. In that situation, your knowledge and observing skill will reach a higher level.

A well-planned observing session means taking the time indoors to decide what objects you want to see, how they should appear through your telescope, and how you can most easily find them. For example, this chapter includes plenty of information on how the recommended targets should appear in your telescope and efficient ways of locating them. Having this information in mind will set your visual expectations so that you will not overlook these celestial gems. A night of observing can pass by before you know it, so it would be best to set aside time for your planning session earlier in the day, or at least an hour or two before sunset.

First sit down with your preferred method of identifying the stars. This could be a planisphere or other all-sky chart, along with a star atlas. Remember if you are using an atlas, unless you are very experienced, it would be prudent to obtain one that comes in one volume and is small enough to manage outside in the elements. (See Chap. 3 for suggestions on a star atlas that will best suit your needs.) Some are laminated to protect against moisture and will show stars and deep sky objects well beyond the naked-eye threshold, typically down to magnitude 8.0 or lower. You may prefer printing out the appropriate star chart from planetarium software. This will lighten your load of things to carry and allow you to focus on the specific areas of the sky that you are interested in.

Even so it would still be smart to also print out a chart that shows the entire sky at one time. As you plot out the best path to take through the sky, imagine each step you will take as if you were already outside. Mentally reviewing this best path (by star pattern recognition) ahead of time gives you a chance to figure out alternate ways of finding an object in case things do not go as planned.

For example, one challenge amateur astronomers have to contend with is a stray cloud or two passing through the sky. As Murphy's Law would have it, the clouds always seem to cross the exact area that we need to view. What if something like that disrupts the original path you chose to take? If you have made a careful study of the constellations, there should be a good alternative or two. In fact, keep in mind that there could be a much brighter object (like a planet) whose transit may be near your intended target, and that would make it much easier to find! So know the sky and be aware of what is going on in it during the evening you plan to observe. There are also web pages, within the companion websites of astronomy magazines, dedicated to keeping readers informed of what is happening each week in the night sky.

In addition to reviewing a star atlas or printed chart, you should take notes and write down what you plan to view. It could be a list of targets by constellation in order from the easiest to the most challenging. Or you may choose to be more elaborate with your notes and write down brief descriptions, directions, and other hints gleaned from your planning session. If you prefer, the use of planning software can also be very helpful. The free program *Stellarium* is simple and easy to use. In addition to offering a good simulation of the night sky, it has a fairly large database of deep sky objects to choose from, and also allows you to zoom in on any target that is entered. It can also provide a view of the sky at any specified day and time, and it can display what the field of view for a particular object will look like through a finderscope, Telrad, and any number of eyepieces that you designate. The program allows you to input and save those user specifications for use at a later time.

The astronomy program *Sky Tools 3* goes much further. It centers around your specific data—observing site, experience, type of telescope, finderscope, eyepieces, and more. After you input that information the program tells you how objects are likely to appear in your 'scope on a given day and time, based on your experience level and observing conditions. It contains a very detailed star charting system that can be tailored to display objects as you prefer them, including how they are labeled (e.g., common name or catalog identification). Another very valuable feature for use in planning your observing session is the capability to toggle the view from star atlas mode to telescope mode. In the telescope mode you are able to see the naked-eye view of a region, the view in your finderscope, as well as the view in the eye-piece of your choice, all on one screen. The field of view for each piece of equipment is automatically calculated and displayed before you on one page. You only need to print it out and have it beside you when you observe (Fig. 4.2).

Among the many other things this program can do is the ability to create an observing list tailored to your location, date and time of observing, and preference for objects. The preferences for objects can be based on catalogs in its database (such as NGC or IC, Double Stars, etc.), or parameters can be set, such as objects brighter than magnitude 12.0, or all variable stars visible in your telescope on a certain night, and only within specified constellations, etc. The capabilities are very extensive and

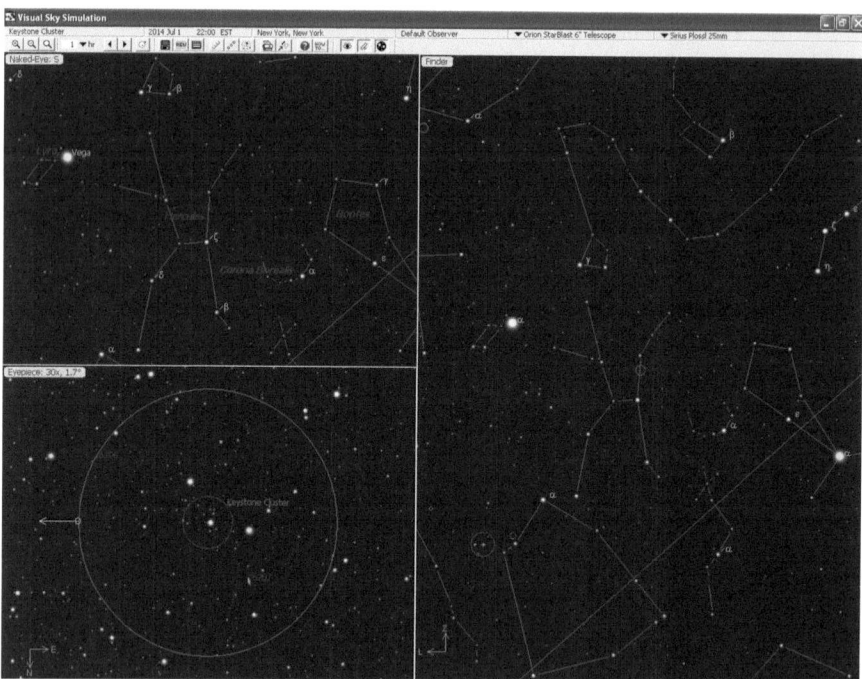

Fig. 4.2 A view of the customizable planning charts within the astronomy software program by Skyhound (Star chart courtesy of *Sky Tools 3* at www.skyhound.com)

beneficial for planning your observing sessions. If you enjoy using a computer, then this or another astronomy software program is well worth looking into. Whether you do your planning with a planisphere and an atlas, or an app and planetarium software, the important thing is that you are able to discover the most efficient ways of finding the objects you wish to see, and then go outside and enjoy the view!

Keep in mind that all deep sky objects are better seen in very dark rural skies as opposed to areas containing light pollution. A description that you read of an object's appearance may very well be based on views under very dark skies. For example, the possibility of detecting the faint central star of a nebula is possible but only with experience under the best observing conditions. If you observe in a location that has moderate light pollution you will need to adjust your expectations of how an object will appear. You will still be able to enjoy a great deal, but the effects of light pollution will be increased (especially in large aperture telescopes) because that is what telescopes do—they gather light, including the light from unwanted sources. A partial solution is to shield your observing area as much as possible from an extraneous light sources. Also, as highlighted previously, a broadband filter should help reduce light pollution caused by sodium and mercury light sources.

A combination of other factors also go into determining how easy or challenging it will be to find a particular object. For one, an object that is dimmer (a higher magnitude number) will be more challenging to locate than ones that are brighter (a lower magnitude number). However, if the size of an object is rather large, then you may find it easier to spot although it may not be very bright. An example of this is the Trifid Nebula (M20). On the other hand, some targets can appear so bright that they are readily seen even though they are very small.

Surface brightness is another factor. This is the overall magnitude of an object spread out over its apparent size. For example, a galaxy may be very large in size, but its overall brightness may not be high enough in relation to its size for it to be easily seen. Pay careful attention to how objects are described so that you have a good idea of what to expect. Please review Chap. 2, "The Best Viewing: Where, When, and How," for a detailed discussion of observing conditions that will affect how well your telescope may perform from different locations.

Remember, taking a few minutes to study the star patterns and to plan what you will observe will go a long way in making your session a success. The extra time spent in advance will save you a lot of time later. It will also make it easier to recognize the constellations, and then to be able to find objects of interest. Planning ahead will also put you in a good position to experience great views of the deep sky that you were not expecting to see.

The Eye and Dark Adaptation

If you have ever attended a star party no doubt you have noticed how serious everyone is about making sure that minimal light is used within the observing area. In fact, the only lighting allowed there is red light—a rule that you also should follow wherever or whenever you observe. Although all light affects the eye's adaptation to the dark, red light has been shown to have the least effect during astronomical observing. Most red lights designed for use by amateur astronomers are adjustable, allowing further control over the amount of light used during observation.

The human eye contains a complex system that focuses light and sends it to the back of the eye in the retina. Within the retina are both cone cells and rod cells. These cells transmit light to the brain as electrical signals, and then the brain interprets them as visual images. The part that we are interested in as related to seeing at night is a pigment called *rhodopsin*, which is made up of vitamin A and proteins. Rhodopsin enables the rod cells to function properly, giving us the ability to see in the dark when our pupils dilate. However, although it takes a short amount of time for the eyes to adjust to bright light, the opposite is true when adjusting to dim light. The adjustment time from exposure to bright light until complete adaptation to the dark takes at least 20–30 min, with a little further enhancement possible during the next hour or so. The reason for this is that bright light bleaches out the structure of vitamin A and the light-absorbing color pigments within the rhodopsin. The vitamin A leaves the rod cells and must then regenerate itself in a different part of the retina called the retinal pigmented epithelium. After some time it regains its proper structure, returns to the rods, unites with proteins, and forms fresh pigments.

This background knowledge makes it apparent just how important dark adaptation will be to your success in observing deep sky objects. Some details you observe will be readily visible, but many features will become noticeable over time, as the sensitivity of your rod cells improve. If the object is bright, such a planet, you will see more color because in that case your cone cells will be the primary receptors. On the other hand, if the object is dim, your rod cells will be the dominate factor in your perception. Although the rod cells will allow your eyes to see a faint or dim object, the colors that your eyes perceive will likely be varying shades of gray.

Averted Vision

When you begin touring the galaxy with binoculars or a telescope, it will at times seem that certain stars are popping in and out of view, as if they are peeking out from behind a dark curtain and hiding themselves again. Some of these effects are related to averted vision, which is a natural occurrence in our eyes when gazing at the stars with optical aid. As our eyes move from one object to the next, certain more sensitive parts of our eyes pick up on details that we did not sense while making a direct inspection of the object. If this can happen without conscience effort, then it stands to reason that we can make a deliberate effort to see more details by the use of averted vision. This, in fact, is a basic skill that every amateur astronomer must become adept at using.

When our eyes focus on a bright image, the light travels to the back of the eye in the retina. There it is absorbed primarily by cone cells adapted for sharp vision and color. This happens in a round area called the macula, located within the retina's center. As light rays are focused, they converge inside of the fovea centralis—a very tiny pit right inside of the macula. When you look directly at something, this is the part of the retina that registers the best image. Three types of pigment in the cone cells allow us to see sharp color images in bright light. The names of these pigments are cyanolabe, chlorolabe, and erythrolabe. Each of them absorbs blue, green, and red light, respectively. Together they enable the human eye to discern in excess of 200 colors.

On the other hand, as explained above, the rod cells are full of a pigment called rhodopsin. This makes the rod cells more sensitive to the dark, and provides us with night vision. However, most of the rod cells are concentrated in the outer part of the retina, which gives us peripheral vision. Therefore, to benefit from the capability of the rod cells, we have to look a little away from those faint astronomical objects, averting our vision, placing them in the area of the eye that contains more rod cells; this provides us with the best chance of seeing dim light. This area is about 20° off the center of vision, but once we begin averting our view beyond this, we lose the sensitivity of this effect. The point of peak sensitivity for most people is between 8° and 16° away from the center of vision.

When averting your gaze, be sure to look in the direction of whichever eye you are using at the eyepiece of your telescope. So if you are viewing with your right eye, avert your gaze to the right. If you are using your left eye, look away towards the left. Looking in the wrong direction can cause the image to fall inside of the eye's blind spot. So to make effective use of averted vision when using binoculars where both eyes are open, try averting your gaze upward. This places the image just below the center of vision while also avoiding the blind spot. Although not as effective as a horizontal shift, it should still prove useful. Vision in everyone is different, and so you will need to experiment with averted vision for yourself in order to find the method that will yield the best results.

As you look at an image, be patient and allow your eyes several moments to take in the view. Make sure that your body and eyes are as relaxed as possible because this will help improve your ability to detect more stars and faint objects. Make sure to eat nutritious food, stay hydrated, and breathe in deep so that you have plenty of oxygen.

Star Clusters

First, let us gain a brief understanding of what star clusters are. There are two types of star clusters: open and globular. Astronomers have found more than 1,000 open clusters and as many as 20,000 may inhabit our Milky Way Galaxy. Most of these clusters have at least 20, to as many as 1,000 stars, and span a distance of about 10–30 light years. As the name suggests, they appear as an open or loose collection of stars, and when viewed through binoculars or a telescope are very beautiful.

On the other hand globular star clusters have a definite spherical shape, and the stars contained inside are packed tight. Compared to open clusters, globular clusters are very old. Many of them have existed for billions of years. They contain between

10,000 and 1 million stars inside a space about 100 light years in diameter. When viewed in a telescope with a wide-field eyepiece, the view is nothing less than spectacular!

Let's now take a tour of some of these exciting objects. The principles that you learn here can be applied during any time of the year. The bright light of the Moon will often prevent a clear view of stars and other objects, so select an evening for observing that is close to the new Moon. You can easily determine the phase of the Moon by consulting a current magazine that deals with astronomy, or by going to the Internet and typing "Moon phases" in the search engine. Another option would be to check your monthly wall or pocket calendar, since many of these include the Moon phases as a common feature. All of the clusters mentioned in this section are easily viewed in a telescope of 4 in. (102 mm) or more. You may find that you prefer viewing some of these objects in binoculars; therefore be sure to alternate your observations between both your telescope and binoculars, and judge for yourself.

Globular Star Clusters

Let us begin with what many consider to be one of the best globular clusters in the summer sky—**M13** in Hercules (NGC 6341), about 26,000 light years from Earth. In very dark skies, you may be able to just glimpse this fuzzy ball with your naked eyes. You will recall that Hercules lies midway between the stars Vega and Arcturus. Using binoculars, find the Keystone asterism in Hercules and center its brightest star Zeta (ζ) Herculis in your field of view. Sweep north towards the next corner star Eta (ή) Herculis, and about two-thirds of the way there your eyes should catch sight of a hazy patch of starlight. Go to your telescope and sight its Telrad, red dot finder, or finderscope on Zeta (ζ) Herculis. Make the same sweep as you did before with binoculars, approximating the distance as you move your 'scope towards Eta (ή) Herculis.

In at least a 4-in. (102 mm) 'scope the cluster will show up nicely, a fairly round looking bright ball of stars. It looks great in a modest-size telescope of 6–8 in. (152–203 mm) in diameter; hundreds of stars will snap into view, making it one of the best clusters you can observe. Use higher magnification to look for details in this globular. An interesting feature you may see near the center is the convergence of three dark lanes where many of the stars seem to have disappeared. A 10-in. (254 mm) or larger 'scope provides a really stunning view. The cluster appears significantly brighter and the stars seem innumerable (Fig. 4.3).

As the saying goes however, "Beauty is in the eye of the beholder," and so you may find that you have a different favorite. One such contender is the globular cluster **M92** (NGC 6341), also located in Hercules, and about 28,000 light years from Earth. It is about 11 billion years old according to a study published in 2010 by A. DiCecco, et al. [1]. To find it, center your red dot finder or finderscope on Eta (ή) Herculis and move about 8° east to another corner star of the Keystone, Pi (π) Herculis. Our target forms a triangle with these two stars. Move back west about

Fig. 4.3 The Hercules Cluster (M13) was first discovered by Edmund Halley in 1714. Astronomers estimate that there are at least 500,000 stars in this cluster (Image by the author)

halfway (4°) so that you are right between Eta (ή) and Pi (π) Herculis. Now move north about 6°, and you will see the stunning globular cluster, M92. It is a little smaller than M13, but has a bright core and tends to appear more crisp and sharp than its neighbor. M92 is well within reach of small telescopes, but you will need at least a 6-in. (152 mm) to appreciate how well it competes with M13 (Fig. 4.4).

Although both of these clusters lie a similar distance away from us, M13 has a very close neighbor (in angular distance) that is actually quite a bit further away. The small galaxy NGC 6207 (see Fig. 5.3 later) is just under 0.5° to the northeast of globular cluster M13. Yet, this spiral galaxy is millions of light years distant! In fact, according to a study of supernova SN 2004A by M.A. Hendry et al. (2006), NGC 6207 is about 65 million light years from us (20.3 megaparsecs) [2]. At magnitude 11.6, this faint object will be a challenge from light-polluted skies. A 6-in. (152 mm) or larger 'scope will pick it up easily under dark, transparent skies. You may need to move the globular slightly out of the your field of view to be sure its light does not cause you to miss the galaxy. Although NGC 6207 is a spiral galaxy, it will likely appear elliptical in 'scopes under 14 in. Finding this object is a good example of taking time to see other nearby sights after you find your original target. Figure 4.5a, b presents a broad and enlarged view of the area near NGC 6207.

The fall months bring in their own collection of globular clusters. In early evening, high in the sky near the star Enif in Pegasus, are two easy-to-find clusters—**M15** (NGC 7078) and **M2** (NGC 7089). Use Fig. 4.6a to help you locate them.

Fig. 4.4 M92—also in Hercules—is slightly smaller than M13, and is estimated to have about 400,000 stars (Image by the author)

The star Enif means "nose," and reminds us of the nose on the Winged Horse of Pegasus. It is a bright 2.4 magnitude star containing a little orange tint, and lies about 600 light years away. To find it, begin by facing towards the southern horizon. High in the southern sky you should see the Great Square of Pegasus. The 2.5 magnitude star at its southwestern corner is called Markab, or Alpha (α) Pegasi. Along with four other stars Markab forms the head of the horse, with Enif marking the point of the nose. There is a distance of about 20° between these two stars. Also, since both Markab and Enif are about the same magnitude and the brightest stars in the area, it should be easy to distinguish them as markers on the way to M15 and M2.

Now that you have found Enif take a look in the area with your binoculars. You will see two glowing orbs. M15 is the one that's 4° northwest of Enif, and M2 is the one that's 11° southwest of it. Both objects are about magnitude 6.5. Lock on to M15 with your finder and spend some time toggling between low- and medium-power views through your telescope. Each of these clusters are well resolved in a 6-in. 'scope under medium power. M15 appears slightly more dynamic than M2 and is quite interesting considering it has a planetary nebula (designated Pease 1 or PK 65-27.1) nested within the confines of its more than 100,000 member stars. It is approximately 30,000 light years distant, with some estimates as high as 40,000 light years distant. M2 is thought to be a little further out at 47,000 light years. Both clusters are estimated to be about 12 billion years old (Fig. 4.6b, c).

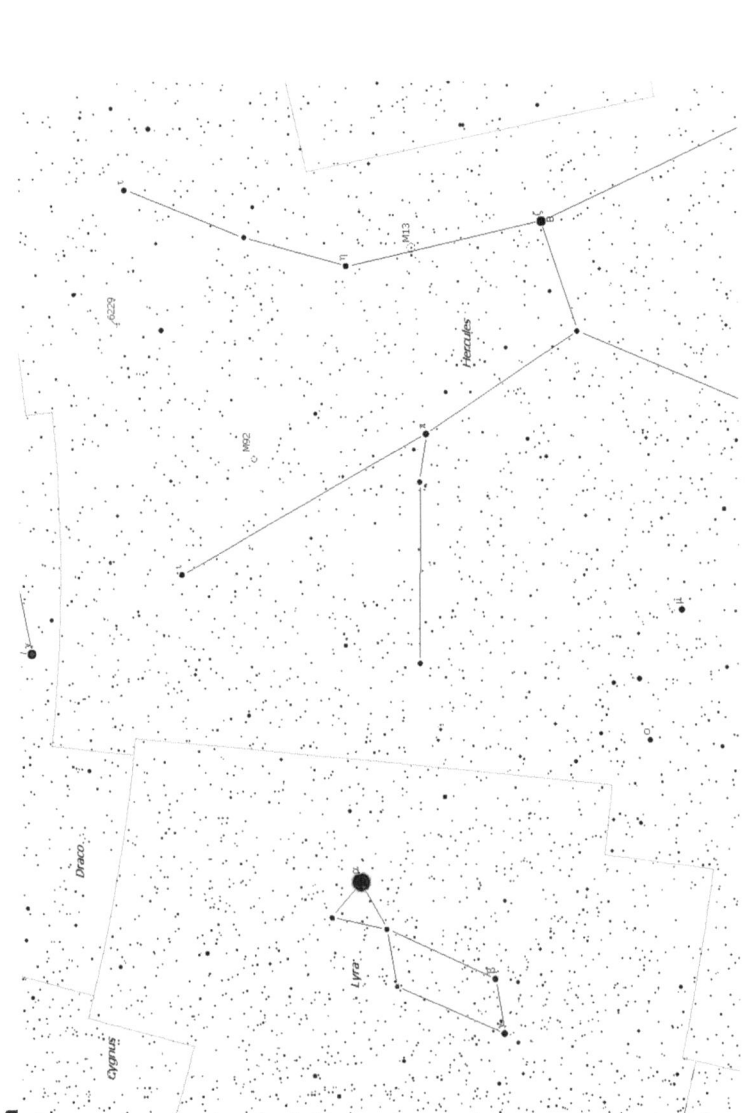

Fig. 4.5 (**a**) This chart displays globular clusters M13 and M92 in Hercules, as well as the smaller, magnitude 9.4 cluster NGC 6229 located 7° to the northeast. (Star chart provided courtesy of *Sky Tools 3* at www.skyhound.com.) (**b**) The spiral galaxy NGC 6207 lies within 0.5° of M13. Recent estimates place it at 65 million light years from us, while M13 is comparatively close to us at 26,000 light years distant. (Star chart provided courtesy of *Sky Tools 3* at www.skyhound.com)

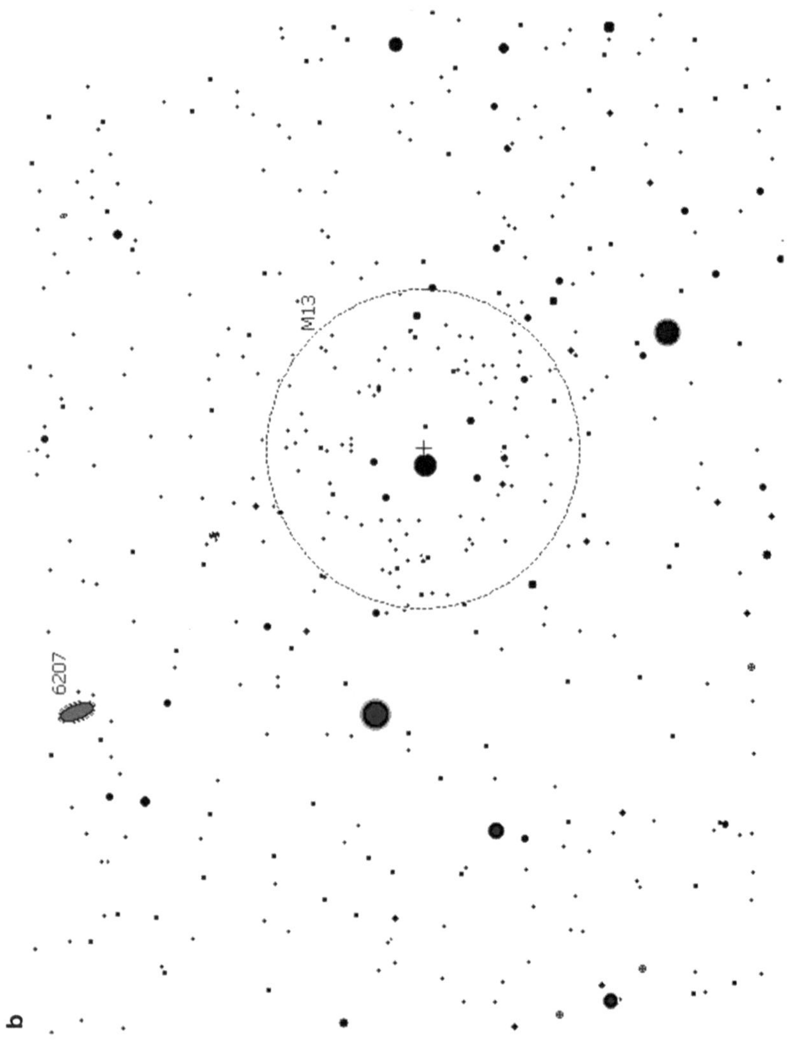

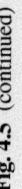

Fig. 4.5 (continued)

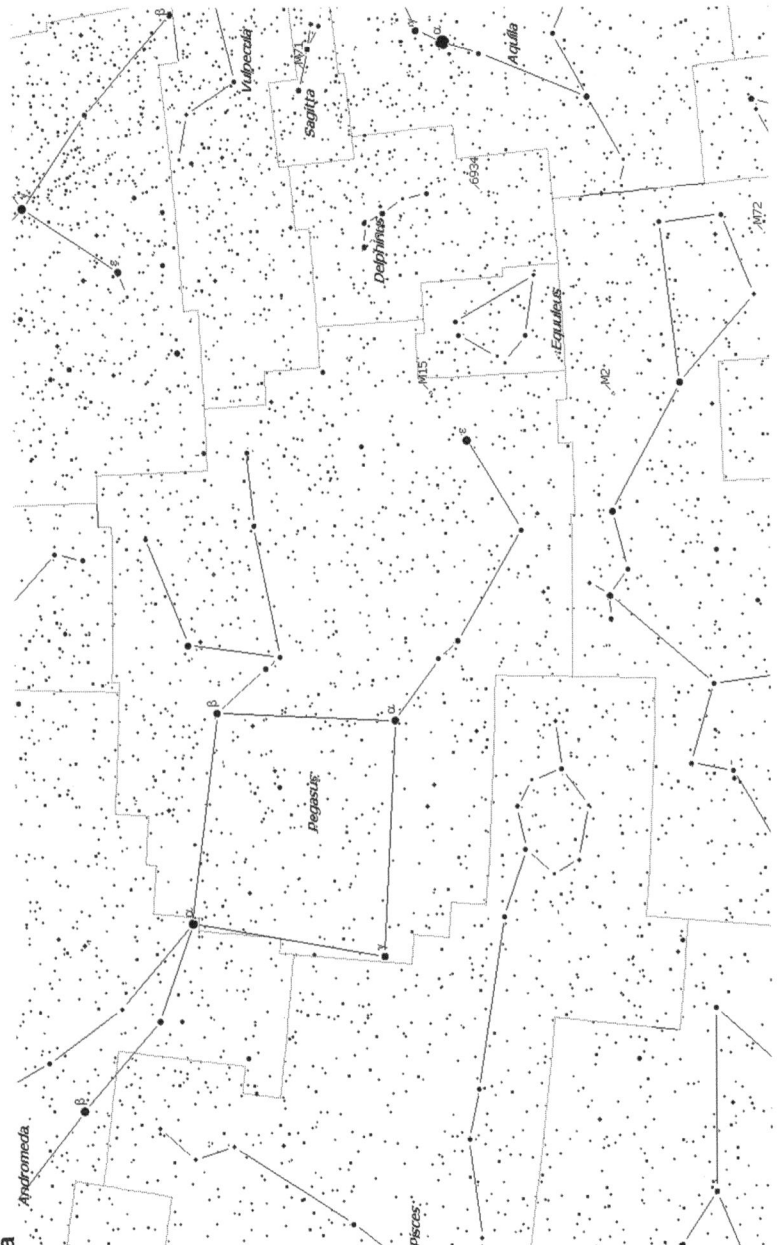

Fig. 4.6 (**a**) In this star chart the globular clusters M15 and M2 are 13° apart and are located near the star Enif (Epsilon [ε] Pegasi). (Star chart provided courtesy of *Sky Tools 3* at www.skyhound.com.) (**b**) M15 in Pegasus. This image was taken by the Kitt Peak National Observatory 0.9 m telescope in 1998. (Image courtesy of NOAO/AURA/NSF.) (**c**) M2 in Pegasus, about 47,000 light years distant. (Image by the author)

b

Fig. 4.6 (continued)

c

Fig. 4.6 (continued)

What is it about a globular cluster that makes it so beautiful to observe? Is it the intense brightness of a cluster? Maybe it is the eerie glow that some globulars tend to emit. Some have stars that are tightly entangled, whereas others are much looser in concentration. At times dark lanes can cut through a cluster, as some observers report seeing in M13. Color in an object is always desirable, even though it may be limited to green or blue hues in the case of objects such as these. Everyone has certain features that appeal to them. What appeals to you?

With these luminous clusters, you have the opportunity to begin developing the art of averted vision. Later you will explore the use of averted vision in connection with much fainter objects. Which of these two clusters do you prefer while using direct vision? What areas of each cluster become more visible when using averted vision? Is your view enhanced by increasing the magnification, or do you prefer lower magnification and a wider field of view? While searching for these clusters did you stumble across any other interesting stars or objects along the way? Days or weeks later your memory of this observing session may fade, but making a record of it will help you to recall more of the experience. Some like to use a pen and notepad, recording the date, time of observation, equipment used, magnification, weather and seeing conditions, along with a detailed description. Some amateurs like to use a log sheet to record this information. (Please see Chap. 3 for a discussion of this, along with a sample log sheet in Fig. 3.3). Another effective way of recalling your observing session is to use a digital recorder. Digital recorders are both readily available and can be obtained at a modest price. They can provide a convenient way to get all of your thoughts out while not having to stop viewing at the eyepiece. (Be sure to turn off the backlight feature if it has one, otherwise you will ruin the dark adaptation of your eyes!) Later if you choose to, you can make a written record of your observing session, based on your voice memo.

Open Star Clusters

Now we turn our attention to some easy to find open star clusters. The first one we'll consider is well placed in the northern constellation of Cassiopeia during the fall and winter seasons. This is a rich area full of all sorts of deep sky objects, and viewing them this time of year takes advantage of more transparent skies. You will only need to identify and then star pattern jump to two very recognizable groups of stars. To find open cluster **M103** (NGC 581), first locate Polaris and then move 25° east-southeast to the "W" asterism of Cassiopeia; on the way there you should see a group of stars shaped like a "T" lying on its side, but pointing straight towards the 3.4 magnitude star Epsilon (ε) Cassiopeiae. Follow the first leg of the "W" to Delta (δ) Cassiopeiae, and once there move back northeast just 1° and you will see this nice little cluster. It is easy to spot in binoculars, and looks beautiful in a small telescope (Fig. 4.7a, b).

The group of stars contained within M103 are about 20 million years old, span a distance of at least 15 light years, and lies 8,500 light years from Earth. It contains

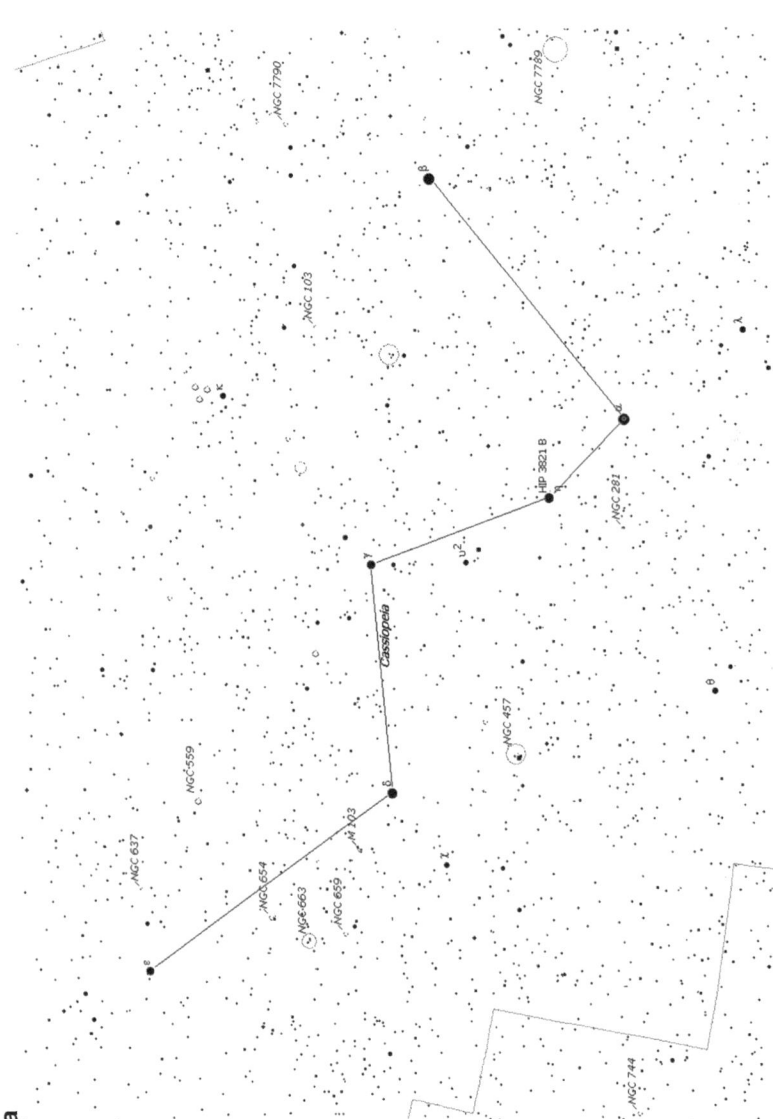

Fig. 4.7 (**a**) This area in Cassiopeia is rich in clusters. (Star chart provided courtesy of *Sky Tools 3* at www.skyhound.com.) (**b**) Open Star cluster M103. (Image by the author.) (**c**) The E.T. Cluster in Cassiopeia (NGC 457). (Image by the author)

b

Fig. 4.7 (continued)

Fig. 4.7 (continued)

about 77 stars, and about 21 of them are variable stars. The gem right in the center of the cluster is a red giant star designated SAO 11826, which shines at magnitude 8.5. Although some observers do not notice the color in this star, most do see a clear orange or red color. More than likely you will become aware of this star's striking contrast with its neighbors when you first view the cluster. Although in Fig. 4.7b the color of this star is not apparent, when viewed under dark transparent skies its color should pop clearly into view.

Within just 2.5° of M103 is a little cluster with a lot of character, commonly known as the E.T. Cluster (**NGC 457**). To find it, simply go back to Delta (δ) Cassiopeiae in the first leg of the "W." Move 2° south-southwest, and its familiar shape will come right into view (see Fig. 4.7a). The two brightest members form the eyes of the extraterrestrial, while the others clearly form the arms, legs, and feet. Easily seen in a small telescope, the stars that form one of the arms appears to be raised in a friendly wave.

Figure 4.7c is okay for a preview, but make no mistake about it—the real thing is a spectacular sight, and one that you will not soon forget. Recent studies count more than 100 members of this group that is 25 million years old and is 9,000 light years distant. Like many other open or globular clusters, it contains stars that are much younger than the majority of main sequence stars of this group. Some of those stars are called blue stragglers. A blue straggler is thought to form when two low mass stars in a binary system merge (probably the case within an open cluster), or in the case of very tightly packed globular clusters, the merger can be caused by a dynamic collision of two stars. According to a study by Fitzsimmons in 1993, there is at least one such object within NGC 457 [3].

M41 (NGC 2287) is a 200-million-year-old open cluster of about 100 stars. This group of stars is currently estimated to be 2,300 light years away. Finding this cluster is simple because it is very close to Sirius, which is the brightest star (magnitude −1.47) in the autumn sky and all year round. Sirius is about 21° southeast of Orion's Belt. From there move 4° south. This magnitude 4.5 cluster is an easy target in binoculars and a naked-eye object under good conditions (Fig. 4.8).

As you continue to observe clusters, you will become so attuned to their existence in the sky that you'll hardly need a map to find them. Especially under dark skies, they seem to reveal themselves everywhere you look. Just sweeping the sky with binoculars for clusters (open or globular) and then following up with a close look through the eyepiece can provide hours of great observing. For now let's continue to highlight a few more and determine the easiest way to locate them.

Two of the most beautiful open clusters visible to the naked eye are in Perseus, and are often called the **Double Cluster**. The eastern half of the cluster is **NGC 884** (sometimes referred to as Chi [χ] Persei) and the western half is **NGC 869** (sometimes referred to as h Persei, although originally Chi [χ] was used to denote both clusters in Bayer's *Uranometria*). Recent studies on the pair indicate that their stars were born at the same time and so share a common age of about 13 million years. They are both moving together through the galaxy at a rate of about 22 km per second. The Double Cluster swarms with thousands of stars about 7,500 light years distant. At magnitude 4.4 they are easy naked-eye objects, beautiful in binoculars, and stunning even in small telescopes.

Fig. 4.8 M41 in Canis Major. The bright star in the lower left part of the image is a foreground star not a part of this group of stars. The cluster extends about 25 light years in size. (Image courtesy of NOAO/AURA/NSF)

To locate the glow of the Double Cluster, find the magnitude 2.7 star Ruchbah (Delta [δ] Cassiopeiae), in the "W" asterism of Cassiopeia. From there move approximately 8° southeast, and the cluster is easy to spot, especially in binoculars. A low power, wide field eyepiece will provide a comprehensive stunning look in your telescope, but be sure to try higher powers for a closer look at its stars. Figures 4.9 and 4.10 will give you a good idea of what to expect, but be ready for a much better visual impression when you see it for yourself.

Many have wondered why Charles Messier did not include the Perseus Double Cluster in his famous list of 110 objects. It was clearly marked on an atlas that was well used in his day. However, being a naked-eye object that was observed since antiquity there really was no need for him to mention it, much like many other objects that were easily visible. Although his list grew to much larger proportions than he originally intended, it became clear that it would never expand to the extent of others, such as the list compiled by William Herschel.

Now let's take a look at open cluster **M38** (NGC 1912) in the Auriga constellation. This group of stars is over 200 million years old, and estimates of its distance from Earth range from 2,800 to 4,000 light years. As with most objects, there are various star pattern jumps that you can take en route to your destination (Fig. 4.11a–c).

Fig. 4.9 NGC 884, the eastern half of the Perseus Double Cluster. (Image by the author)

Fig. 4.10 NGC 869, the western half of the Perseus Double Cluster. (Image by the author)

Fig. 4.11 (**a**) True color image of open cluster M38 in the constellation Auriga. (Image courtesy of NOAO/AURA/NSF.) (**b**) The constellation of Auriga is just 22° north of Betelgeuse. (Star chart provided courtesy of *Sky Tools 3* at www.skyhound.com.) (**c**) In this enlarged area near M38 the stars Chi (χ) and Phi (φ) Aurigae frame the stars, forming a triangle asterism on the way to M38. (Star chart provided courtesy of *Sky Tools 3* at www.skyhound.com)

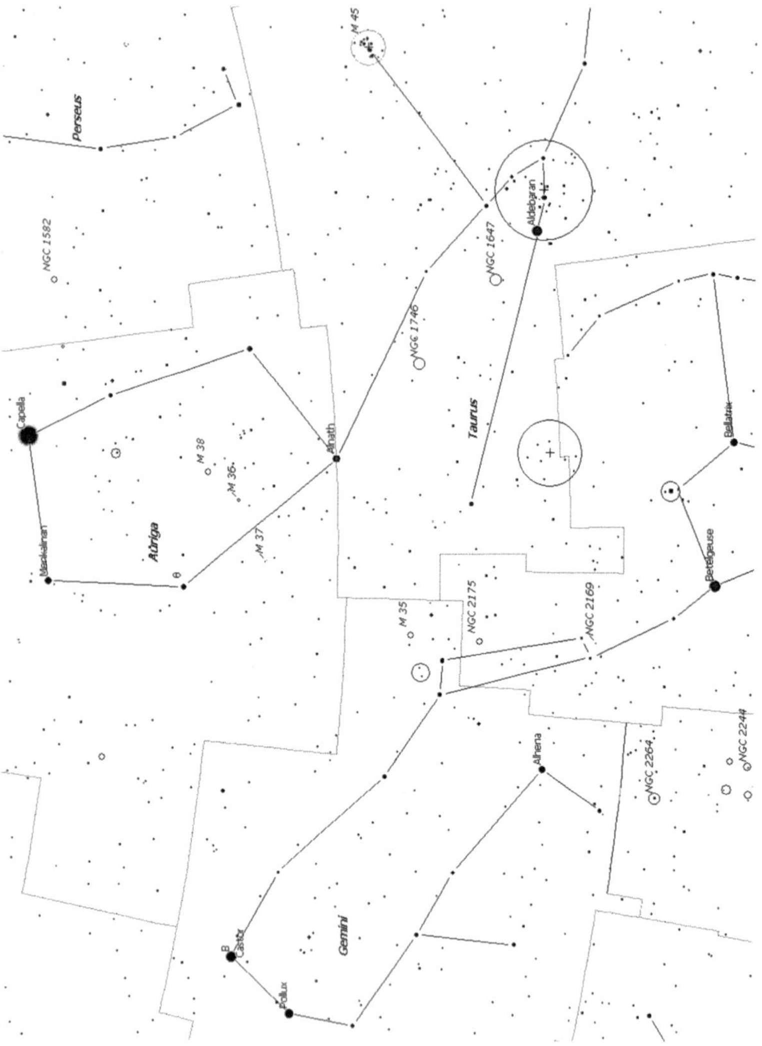

Fig. 4.11 (continued)

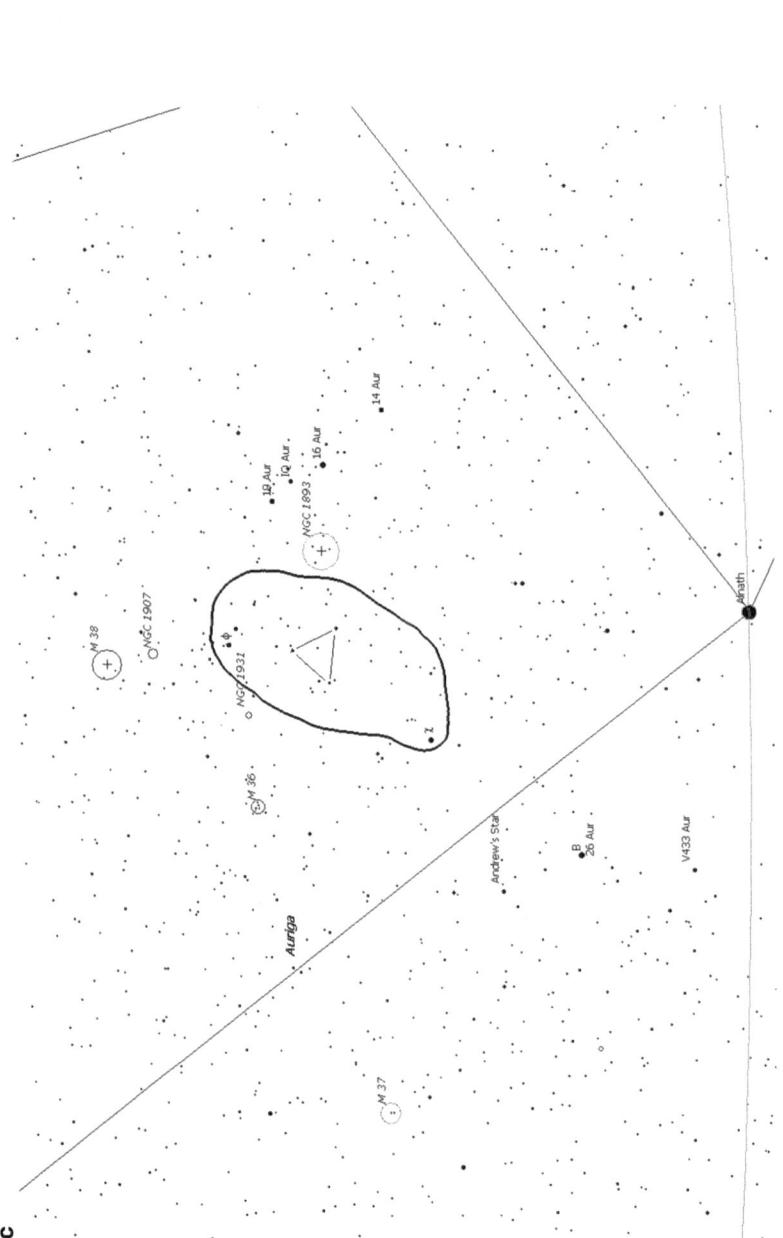

Fig. 4.11 (continued)

Let's first look at the route via Orion. In autumn face east-southeast towards the Orion constellation. About 22° north of Betelgeuse is the 1.6 magnitude star Alnath or Beta (β) Tauri, in Taurus, although it actually forms part of the asterism that represents Auriga the Charioteer. M38 lies about halfway between Beta Tauri and the very bright star Capella in Auriga. On the way from Alnath to M38 is a triangle of two magnitude 6 stars and one that is magnitude 7, that sits between two brighter magnitude 5 stars—Chi (χ) Aurigae on the south side and Phi (φ) Aurigae on the north side. M38 is just over 1° north of that star. In fact, to get to it you will probably cross over the smaller cluster NGC 1907. The two, although near each other, are physically unrelated.

Another very easy way to find the M38 cluster is to first locate the bright star Capella. It is about 43° south of Polaris during autumn and winter months. Once you are at Capella, simply look 10° south to find M38. A beautiful cluster to view, even in small telescopes, but with larger apertures of 6 in. or more, stars are well resolved and some color can be observed. Take your time and explore the area; as the accompanying chart shows, there are several other clusters to be seen nearby. Remember to take brief notes on a digital recorder or observation log, which will help you to remember and assess how each object appears to you.

In our introduction, we discussed the Beehive Cluster (sometimes referred to as the Praesepe Cluster, or NGC 2632, see Fig. 4.1a) or **M44**, a 6.3 magnitude group of about 1,000 stars that are 660 million years old. It spans 17 light years across and is situated about 580 light years from us, visible in the constellation of Cancer. A fascinating study in 2012 found two Jupiter sized planets orbiting Sun-like stars inside this cluster [4]. Undoubtedly in the years to come, many more exoplanets will be discovered orbiting stars that are a part of clusters.

In the spring of each year, you will find this group high overhead in the early nighttime hours. In late March go outside at about 9:00 p.m. and find the two most prominent stars of Gemini, magnitude 1.5 Castor and magnitude 1.0 Pollux, which are only 4.5° apart. Both of these stars are clearly marked on a planisphere or all-sky map. From Pollux, go southeast about 15° and you will find a hazy patch of stars visible to your naked eyes. Center this hazy patch in your red dot finder or finderscope and enjoy. The cluster looks beautiful in binoculars and spectacular through at least a 4-in. (102 mm) telescope using low power.

Another open cluster worth targeting is found in the southern summer sky within the brilliant bands of the Milky Way. It is has been nicknamed the Northern Jewel Box for its apparent resemblance to a box full of shimmering diamonds. You will find **NGC 6231** in the constellation of Scorpius (The Scorpion) visible early in the evening. Begin by facing south and sighting the red, magnitude 1.0 star Antares in binoculars. Then follow the stars down the length of the Scorpion until you arrive at Zeta (ζ) Scorpii. Be sure that you have a printed star chart of this region handy or have the appropriate page in your star atlas open. Binoculars will reveal it to be a double star. One is an orange star, magnitude 3.5, and the other is a blue supergiant star, magnitude 5.0. Now be sure to center this double star in the red dot finder or finderscope and then look through the eyepiece of the telescope. From this pair of stars, arriving at the Jewel Box is simple—move north just 0.5°. This large open cluster is about 5,900 light years distant and contains at least 120 stars. It is also of interest that this cluster is within a spiral arm that is much closer to the middle of our galaxy than the spiral arm that contains our Sun. Later, our review of nebulae will take us to the constellation Sagittarius, where astronomers have located the very center of our Milky Way Galaxy.

This part of the sky is very rich in stars and other visual treasures. It is easy to run into other notable objects, so by all means take your time and explore. We will conclude this brief excursion with a globular cluster that you may have already noticed near Antares, while on your way to NGC 6231, but for the moment let's continue towards the end of the tail in the constellation Scorpius. After consulting your star chart find the star Lambda (λ) Scorpii—a bright star, magnitude 1.5. Scan the area, go about 5° northeast, and you will see a large open cluster. The bright open cluster **M7** is 800 light years distant and has an apparent size that is about four times as large as M13 in Hercules. If you are using binoculars, another open cluster (**M6**) is visible just 4° to the northwest. That cluster is a little further away from us, at about 1,500 light years, and although not as large or as bright as M7, it is well worth viewing through a telescope.

Now let's return to the bright star Antares. Just 1° west of Antares we find **M4**, a nice globular star cluster 10,000 light years away. This cluster is similar to M13 in both physical size and apparent size, although it may not be as spectacular to view. What do you think? Remember to try various magnifications, use averted vision, and record your observations.

There are numerous other open and globular clusters you will want to try and locate. With each success the task will become easier. Here are a few additional targets: **M3**, **M5**, **M36**, **M37**, **NGC 6144**, **NGC 1907**, **NGC 1931**, and **NGC 1893**. Consult your star chart or atlas, grab your binoculars, and begin the hunt!

Helpful Things to Consider

What bright stars will you target first in order to locate these objects? After reviewing a star chart or atlas, are there any asterisms that occur in your mind that you can match to the sky as you scan with your binoculars or finderscope?

If you did not locate your target at first, did you allow time for your eyes to adapt to the dark? Did you try averted vision?

How does your view of the object change with higher magnification? What is your preferred magnification for a particular object?

Always remember that if you do not see your target, start the search again, retrace your steps and slowly make a second sweep, and then if necessary, another one. Eventually you will find what you are looking for. Also do not forget to enjoy sights that you stumble upon along the way. Consult your star atlas or planetarium software so as to identify the objects that you have encountered.

Nebulae

There are several different types of nebulae, but for the purpose of our search for interesting nebulae in this chapter, we will focus on *diffuse nebulae* and *planetary nebulae*. Diffuse nebulae often look like clouds of scattered light. They are stellar clouds of hot gases powered by the energy from the stars inside of them. These nebulae have irregular shapes and are often perceived by our eyes as blue or green. These types of nebulae can reflect light (*reflection nebulae*), emit light (*emission nebulae*), or obscure light, as in the case of dark nebulae (also called *absorption nebulae* or *dark dust clouds*).

A reflection nebula appears blue, as the dust is reflecting more of the blue light. An example of this is the reflected blue light in the Pleiades. An emission nebula is one that has absorbed powerful ultraviolet energy and then emits light in the form of ionized hydrogen, which transmits more red light. An example of one is the Orion Nebula (M42), found below Orion's Belt. The Horsehead Nebula also contains a lot of energy that is emitting this ionized hydrogen (H II). Its famous dark cloud, shaped like the head of a horse, is a good example of a dark cloud obscuring light.

A planetary nebula forms as a result of the turbulent activity of an aging star that has reached the red giant stage. The star pulsates, causing a stellar super-wind that blows material away from the surface. The expelled material forms shells of gas heated by the remaining hot central core of the star. When astronomer William Herschel first viewed planetary nebulae in 1785, he thought they resembled planetary disks, and so the name *planetary* nebula stuck. One well known example of a planetary nebula is the Ring Nebula (M57) in Lyra. Planetary nebulae are not limited to resembling only the disk shape of a planet. As we will see, they form a fascinating array of various shapes and sizes.

Before you set your eyepiece on a nebula, a brief reminder about color imaging is appropriate. You are going to see things that may take your breath away, but do not expect to see deep sky objects in the same vivid colors that you see in pictures. Often the imaging is created in false color, in order to bring out details that we might miss, or to specify a portion that appears in a different wavelength of light. Some true colors become more apparent in long exposures and provide us with a vivid picture of what the object would look like if our eyes were as sensitive as our photographic equipment. In visual astronomy, however, the colors most visible to our eyes are green, blue, and gray. In some cases, a very large amateur 'scope will show a little more color to the eye of an observer than a small 'scope will. In the next few pages, some color photographs are shown, as well as black and white shots that are more likely to be what you will see in your own telescope.

The first stop in our nebulae tour will be the Orion Nebula. This nebula has long been a subject of astrophysical and visual study because of its relative close proximity to Earth and its stunning beauty. It is located in one of the most recognizable constellations in the night sky, second only perhaps to Ursa Major which contains the Big Dipper. It is often recognized by the three stars forming a belt, and also the bright star Betelgeuse on the shoulder of Orion. Betelgeuse is a variable, red super-

Fig. 4.12 The Orion constellation amid the backdrop of the Milky Way. (Image courtesy of ESO/C. Malin [christophmalin.com])

giant star, varying from a minimum magnitude 1.3 to a maximum of 0.40, during a period of 2,110 days. Its angular size has been measured directly by the Hubble Telescope, revealing it to be extremely large—630 times the diameter of our Sun! Betelgeuse is exactly 130 pc away (or about 424 light years) (Fig. 4.12).

The Orion constellation is visible rising in the eastern sky from late November and remains well placed for viewing in the sky through the month of April. Even when viewing this constellation in skies with moderate light pollution, the nebula can be seen with the naked eye. First locate the three stars that form Orion's Belt. Allow your eyes to gaze downward a few degrees below the farthest star on the left of the belt. This area is in the sword hanging from Orion's Belt (Fig. 4.13).

If you have never viewed the Orion Nebula through a telescope before, you are in for a real treat. You will see a glowing patch when looking straight at it, and averted vision will make it a little more visible. Next, aim your binoculars at the glowing patch. You should be able to see a large bluish-green patch of nebulosity in a field of many beautiful stars. In fact, the forces within this nebula are still giving birth to hundreds of new stars. The approximate diameter of the entire nebula is 30 light years, which is 20,000 times the size of our Solar System! As you aim your telescope on this exciting object 1,500 light years away, take your time and discover as much detail as you can (Fig. 4.14).

Using low power will give you a wide field of view, allowing a more comprehensive viewing window on the many stars and wisps of nebulosity, even in moderately light-polluted skies. Through a 6-in. (152 mm) telescope at about 75× it will appear like a light bluish ghost. Although the nebula appears as one object, it has been classified as two separate ones. The smaller round nebula looks somewhat like a head (**M43**) attached to the larger portion, which to some resembles a cape (**M42**

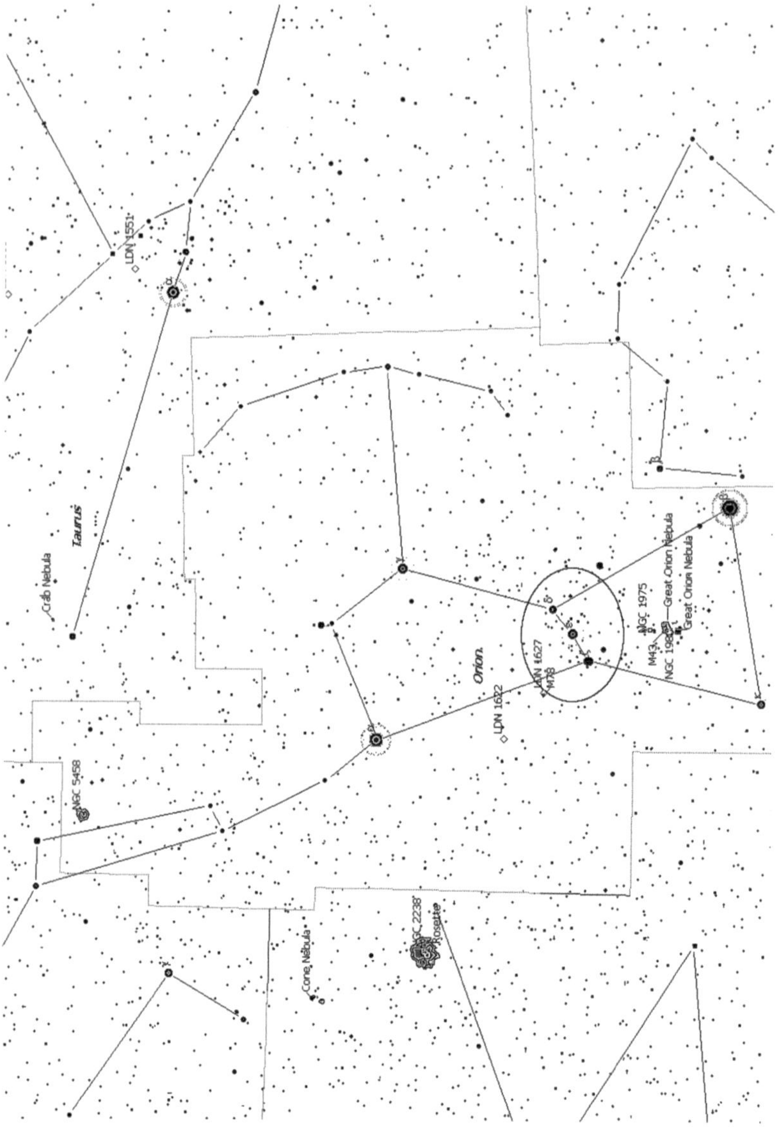

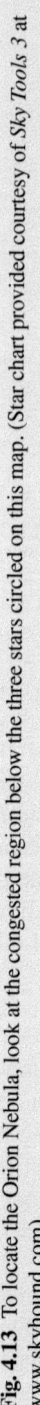

Fig. 4.13 To locate the Orion Nebula, look at the congested region below the three stars circled on this map. (Star chart provided courtesy of *Sky Tools 3* at www.skyhound.com)

Fig. 4.14 The Orion Nebula (M42). (Image by the author)

[NGC 1976]). Under at least medium power make a close examination of the center of this nebula. Inside you should be able to resolve four tiny stars that form the shape of a trapezoid named the Trapezium. Those tightly placed stars are well within reach of a 4-in. (102 mm) 'scope. If you have trouble discerning the Trapezium at first, try using higher power.

At an average age of 300,000 years, these stars are babies compared to our Sun, which is 5 billion years old. Within many parts of the Orion Nebula, astronomers continue to find a lot of activity that is giving birth to new stars. In dark skies with a large telescope of 10 in. (254 mm) or more, the large size of this nebula is revealed and looks stunning. The apparent resemblance of the nebula's shape and color are much more defined, along with a much brighter and more copious star field.

The Orion Nebula is a great place to begin observing nebulae, as well as the perfect way to introduce someone else (such as your neighbor, child, or significant other) to astronomy. It is a showstopper, and most people after seeing it will understand why you find astronomical observing so fascinating. Getting involved with other people and sharing your passion for the observable universe is also a great way to reinforce what you are learning.

An unusual planetary nebula is **NGC 2438**, because it visually appears to lie within the 300-million-year-old open cluster **M46** (NGC 2437). Actually the planetary is somewhat closer to us, at a distance of 2,900 light years, while the open cluster is further out at 3,200 light years.

There have been many studies to determine whether or not the planetary is physically associated with the M46 open cluster itself. There is still not complete agreement on this topic among astronomers. A 1996 study by Pauls and Kohoutek demonstrated that there were similar velocities for both the nebula and some stars within the cluster. However a recent study by Kiss et al. [5] examined many more stars. After examining the radial velocity of 586 stars inside the cluster, along with data from the Spitzer Infrared Array Camera, it concluded that the significant difference in the velocity of the two objects means that they are not physically intertwined. The study does, however, leave open the possibility that the two objects may have been physically associated in the past, but this new data effectively rules out any present physical association (Fig. 4.15a–c).

NGC 2438 and the open cluster M46 that is its background are easy to find. Locate the star Sirius again as earlier described for locating M41. The cluster and planetary duo are about 14° east. A map for this object is provided in Fig. 4.15c to avoid any confusion, since there are several other objects nearby. M46 is a rich star cluster, but the real treat is to identify the planetary nebula within. The nebula is well within reach of a 6-in. 'scope under average observing conditions, and within smaller telescopes in dark transparent skies. Use medium power and slowly sweep your eyes throughout the array of stars until it stands out. A large 'scope of 10 in. (254 mm) or more will resolve the nebula into its donut shape. This is definitely one of the best objects to observe.

Another autumn planetary is the astonishing magnitude 9 planetary nebula, the Blue Snowball, **NGC 7662**, a great object to view due to its high surface brightness. At a distance of about 4,000 light years from us, this nebula's central star is a hot blue (110,000 K) 13.2 magnitude star, surrounded by a triple shell. The material in this nebula is expanding at a rate of 26 km per second, and the object itself is speeding towards us at a rate of 13 km per second. A bluish or blue-green color in this planetary is quite obvious in a 10-in. (254 mm) or larger 'scope, and the view is an extraordinary display. Although some observers report seeing color in this nebula with a small telescope, usually at least a 6-in. 'scope is required to see the eerie blue disk (Fig. 4.16a).

To locate the Blue Snowball first find Beta (β) Pegasi in the Great Square of Pegasus. About 14° due north look for a 3.5 magnitude star named Omicron (o) Andromedae. The Blue Snowball is about 4° west of that star, just about 2.5° before you arrive at a triad of 4th magnitude stars (ι Andromedae, κ Andromedae, and λ Andromedae) (Fig. 4.16b).

The Dumbbell Nebula, or **M27** (NGC 6853), is a must-see object located 1,300 light years away in the constellation of Vulpecula. Its central star (shown in Fig. 4.17) is a hot (100,000 K) white dwarf, magnitude 14.0. The nebula spans a distance of around 2 light years and is thought to be 20,000 years old. This planetary is fairly easy to spot in binoculars. Start with the 3rd magnitude star Albireo (Beta [β] Cygni) in Cygnus. Visualize an 8° long line moving southeast that extends from Albireo to 13 Vulpeculae (magnitude 4.5) and finally ends at 14 Vulpeculae, which is about magnitude 5.5. The Dumbbell is less than ½° south of that star. This is an easy object for small telescopes, but in a medium 'scope or larger much more detail can be seen.

a

Fig. 4.15 (**a**) Image of the planetary nebula NGC 2438 within open cluster M46. (Image by the author.) (**b**) Enlarged image of planetary nebula NGC 2438. (Image by the author.) (**c**) Planetary NGC 2438 is just 14° east of Sirius inside the boundaries of open cluster M46, indicated by the arrow. (Star chart provided courtesy of *Sky Tools 3* at www.skyhound.com)

b

Fig. 4.15 (continued)

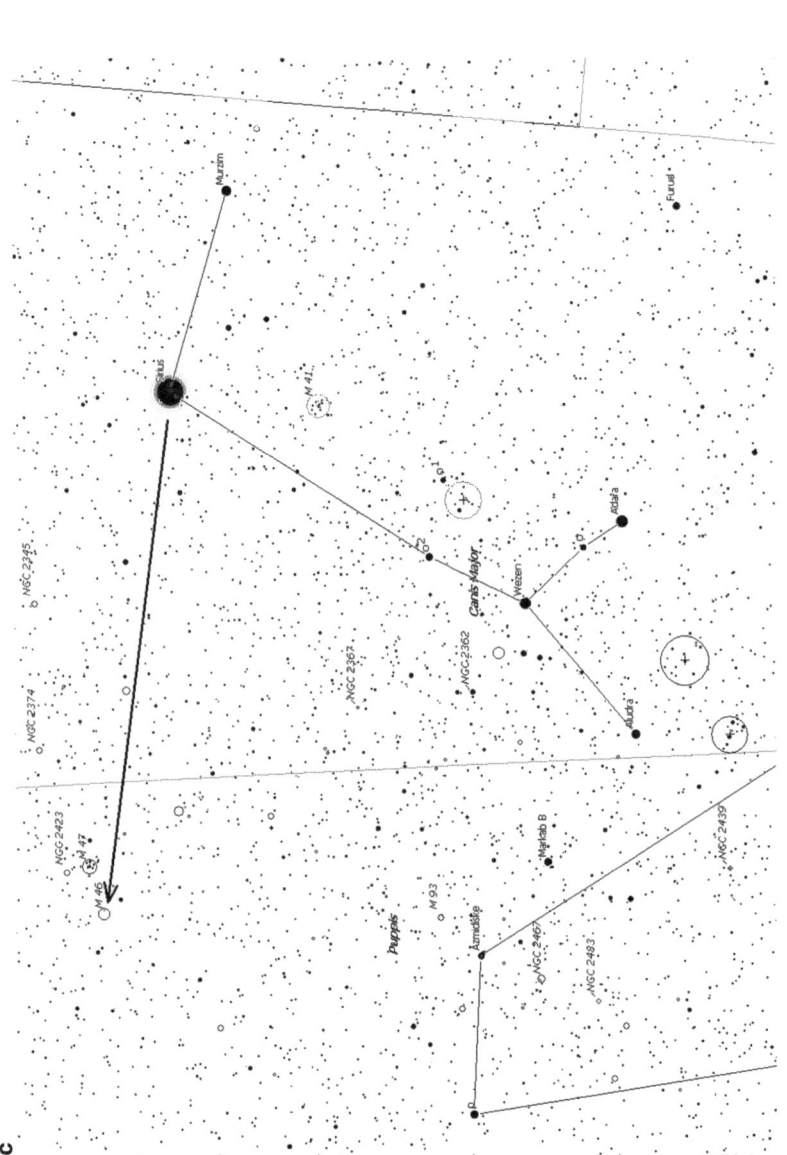

Fig. 4.15 (continued)

a

Fig. 4.16 (a) The Blue Snowball (NGC 7662). (Image by the author.) **(b)** The Blue Snowball (NGC 7662) appears between the stars o Andromedae and ι Andromedae. (Star chart courtesy of *Sky Tools 3* at www.skyhound.com)

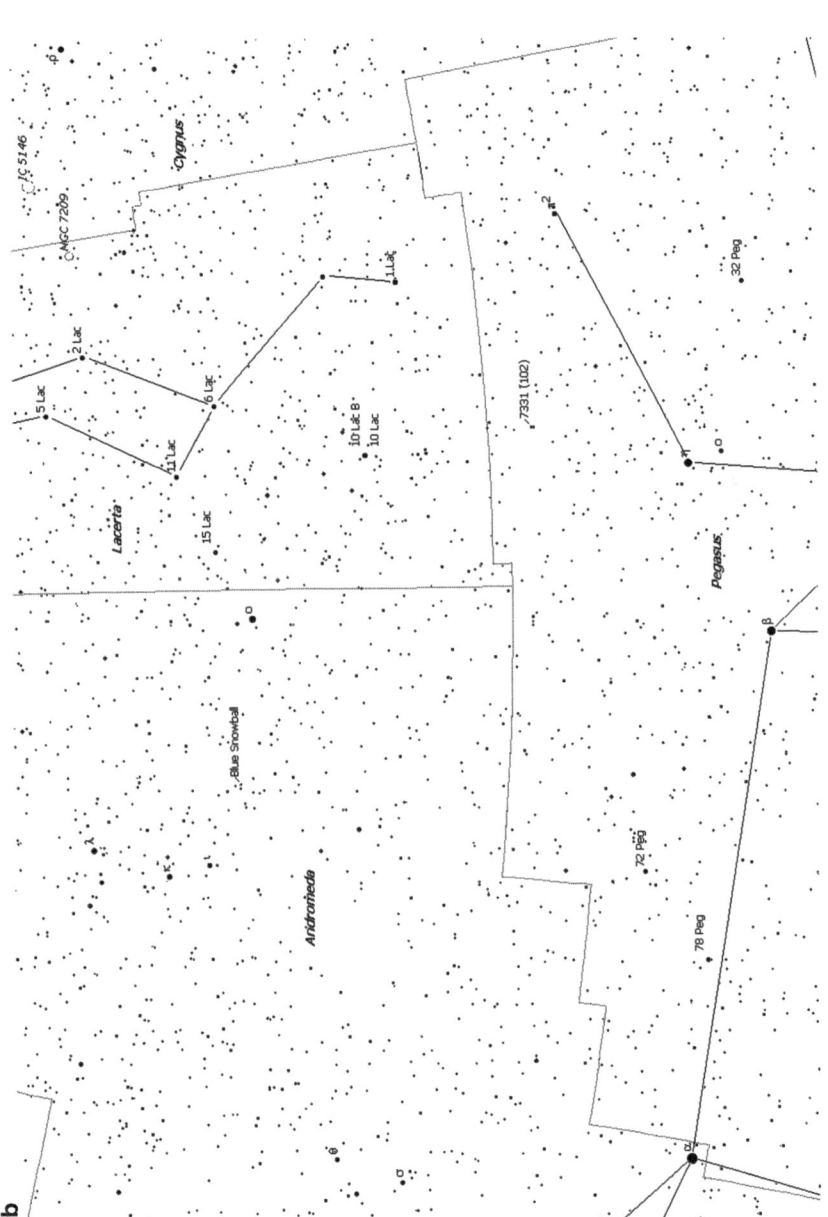

Fig. 4.16 (continued)

Fig. 4.17 The Dumbbell Nebula (M27) in Vulpecula. The 14th magnitude white dwarf is visible right in the center of the nebula. (Image by the author)

The planetary nebula **M76** is called the Little Dumbbell Nebula, because its shape is so reminiscent of M27. You can find this little gem in the constellation of Perseus. Using binoculars, find Delta (δ) Cassiopeiae, and within your field of view to the southeast should be a pentagon-shaped group of stars marked prominently by the brighter Chi (χ) Cassiopeiae. Now move due south about 7° to the magnitude 10 planetary. You may wish to put the 4th magnitude star Phi (Φ) Persei in your finderscope first. Then, using an eyepiece with at least a 1° field of view will put Φ Persei in the same field as the nebula. Move the star just out of the eyepiece's field of view to the south so that it does not interfere with your detecting it. As with M27, the Little Dumbbell can be seen in small telescopes (Fig. 4.18).

At first M76 looks light a bright bar, but careful examination with averted vision reveals a shape that mimics the Dumbbell Nebula. A 6-in. (152 mm) telescope or larger will make this little nebula very well pronounced, clearly showing each half of the miniature dumbbell. You may need to use a narrowband or O-III filter to make it stand out under very low power. However, upon moving up to around 70–100×, this planetary becomes much more obvious, and the view without a filter is better. The thing about filters is that they provide a different view to each individual. On this object specifically, an observer situated right next to you may really appreciate the enhancement, while you may see no significant improvement. Sometimes aperture can be the deciding factor, so if your 'scope is a little smaller than 6 in., using a filter may be the preferred method of making its details more evident.

Fig. 4.18 The Little Dumbbell Nebula (M76) in Perseus. (Image by the author)

The Helix Nebula (**NGC 7293**) is a large planetary nebula that is somewhat more challenging to spot. It is magnitude 7.3 but with a diameter of about 17′ (17 arc minutes), it has a low surface brightness. However it can be seen in small telescopes and responds well in a wide field eyepiece. Medium 'scopes will begin to show a more detailed ring structure, especially with averted vision. Seeing greater detail in this planetary requires that you observe it under very dark skies, and you may also find an O-III filter helpful as well. This nebula is relatively close to us at a distance of 650–700 light years according to the most recent measurements. Its central star is a hot white dwarf with a temperature of about 120,000 K.

You should be able to glimpse it in binoculars that are 10×50 mm or more, with a keen eye and a dark, transparent sky. First go back to Enif, where we began to hunt for globular star clusters M15 and M2. About 11° southeast of Enif is the star Alpha (α) Aquarii (magnitude 3). The Helix Nebula is about 21° south of Alpha (α) Aquarii, and sits between Delta (δ) Aquarii and Delta (δ) Capricorni. These stars with their Bayer designations are shown on both the broad map of the area and the enlarged map. The numbers beside the stars on these maps are the magnitudes, not the Flamsteed numbers. Be sure to add one decimal place to the number shown, so that Delta (δ) Capricorni, which is labeled as "δ(28)" on the map, becomes magnitude 2.8. Now look at the enlarged map of the area near the Helix Nebula. The nebula is about 3.5° from the 4.7 magnitude star 66 Aquarii, and also about 3.5° from the 4.1 magnitude star 21 Aquarii (Fig. 4.19a, b).

Another way of looking at the area is that the Helix sits between two triangles formed by stars of similar magnitude and is framed within two stars on the outside

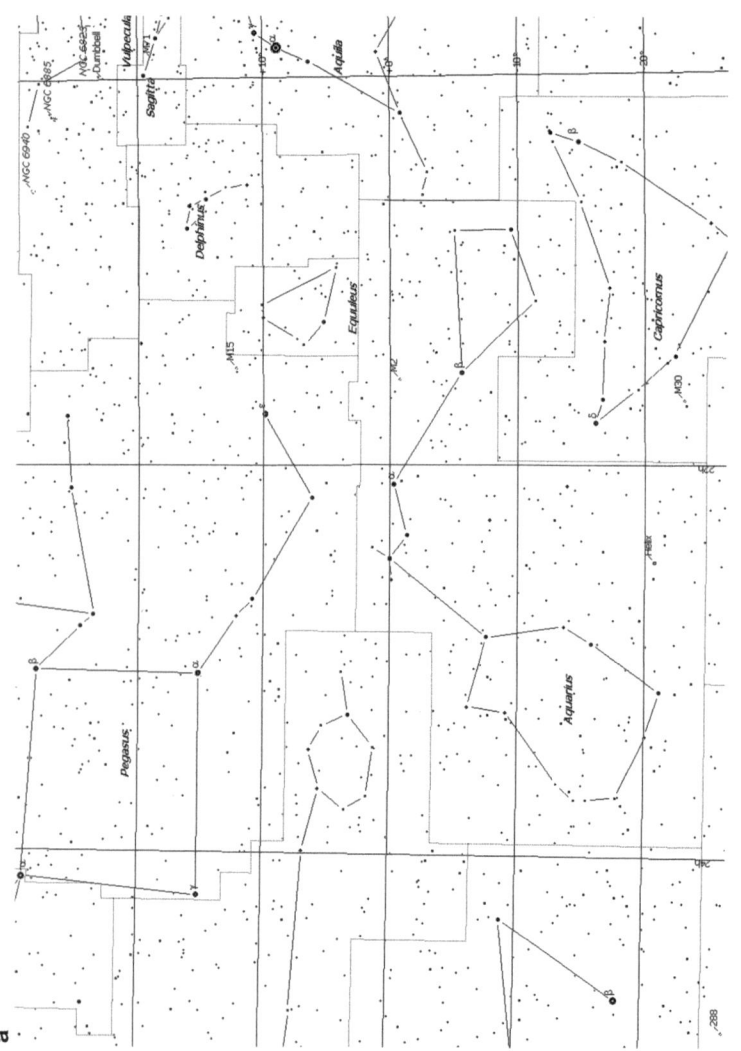

Fig. 4.19 (**a**) Using this map to find the Helix Nebula, follow the path from Enif south for 11° to α Aquarii, and then continue another 21° south. (Star chart courtesy of *Sky Tools 3* at www.skyhound.com.) (**b**) In this enlarged map, the Helix Nebula sits almost in the middle of a 16° long line between δ Aquarii and δ Capricorni. Two triangles surround the nebula within a space of 9° from 68 Aquarii to 35 Aquarii. (Star chart courtesy of *Sky Tools 3* at www.sky-hound.com)

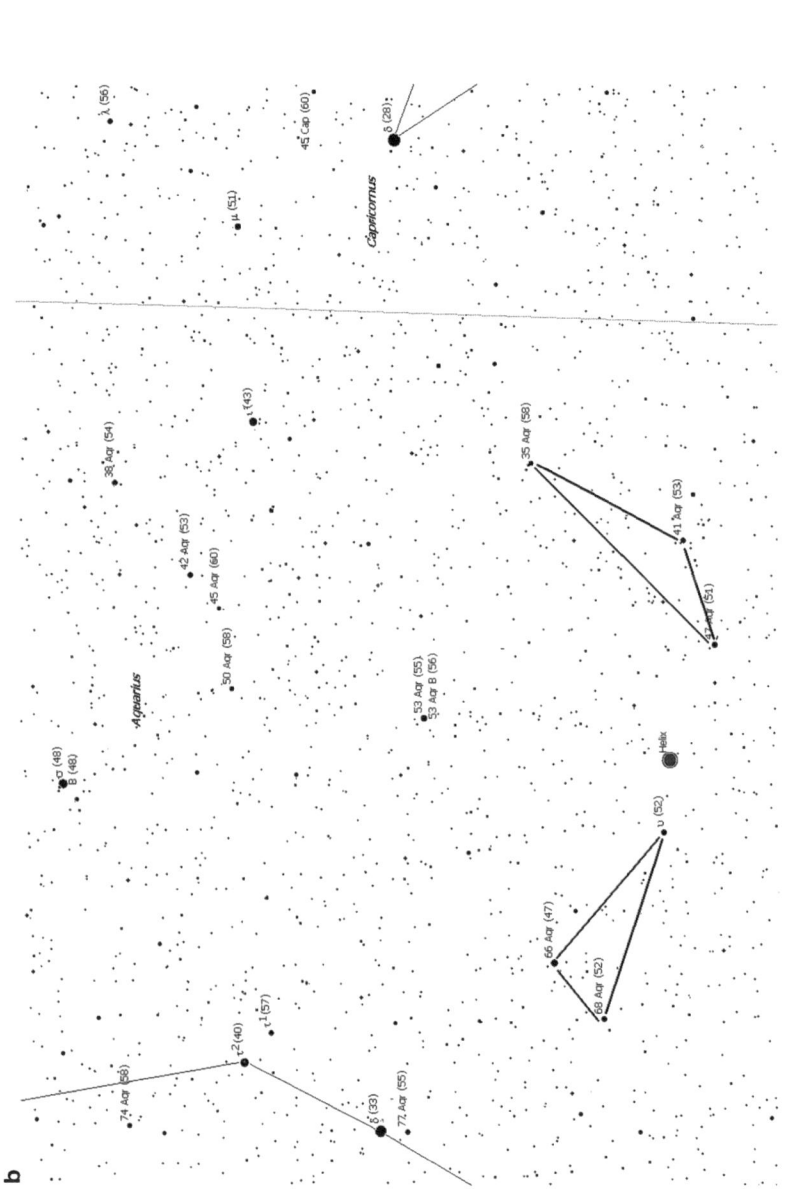

Fig. 4.19 (continued)

Fig. 4.20 The Cocoon Nebula (IC 5146) in Cygnus. (Image by the author)

that are much brighter. (Both Delta (δ) Aquarii and Delta (δ) Capricorni are about magnitude 3 and are 16° apart.)

The Cocoon Nebula (**IC 5146**) is our next quarry, about 3,000 light years away in the constellation of Cygnus. Astronomers have detected within this nebula approximately 100 stars, all part of an open cluster (the cluster is referred to as Collinder 470) that is probably over 1 million years old. You'll find these young suns burning within their cocoon just 4° northeast of 4th magnitude Rho (ρ) Cygni. IC 5146 sits on a line about two-thirds of the way from the bright star Deneb, moving east towards the star designated 2 Lacertae (which is magnitude 4.5). The nebula is detectable in a small telescope, and although it is surrounded by a vast star field, very few stars that are members of the cluster inside are visible in amateur telescopes. A 10-in. or larger telescope should reveal its characteristic round shape and at least one of its dust lanes. Review the map in Fig. 4.16b to help locate IC 5146, which can be found in the upper right hand corner near Rho (ρ) Cygni (Fig. 4.20).

Now let's return to summertime observing to find nebulae in the constellation of Sagittarius, which is home to our galactic center. We are not able to view the galactic center visually, but it has been detected by astronomers using instruments that detect other wavelengths of light. Like Scorpius, the constellation Sagittarius sits right inside of the glowing band of the Milky Way. If you are unfamiliar with the region, the best way to get there is to face south and locate the bright star Antares again in Scorpius. This star denotes the head of the Scorpion. As your eyes follow

Fig. 4.21 The Lagoon Nebula (M8) glows with the light of ionized hydrogen. The image was taken by the Kitt Peak 4-m Mayall telescope in 1973. (Image courtesy of NOAO/AURA/NSF)

the path of stars south, the natural progression flows around to the tail of the Scorpion, ending on the star Shaula (Lambda [λ] Scorpii).

From Shaula look east and observe that it points directly into the teapot asterism of Sagittarius. Imagine a 10° line connecting Shaula with the bright 2nd magnitude star Epsilon (ε) Sagittarii. On the way there you should pass the open cluster M7 again, easily seen in binoculars, although it will be a little north of the line between those two stars. Once you are at Epsilon, it is a short 11° track north to our first target, the Lagoon Nebula. Although observed by Charles Messier in 1764 and entered in his catalog as M8, it was discovered earlier by astronomer John Flamsteed in 1680. It was Flamsteed who devised the numbering system for the brightest stars in each constellation.

The Lagoon Nebula **M8** (NGC 6523), may fast become one of your favorites. It is larger than the Orion Nebula, and just as captivating. With at least a 4-in. (102 mm) 'scope a large dust lane can be seen cutting through the nebula. With a medium-size telescope of at least 6 in. (152 mm), this feature becomes more prominent. On its eastern side an open cluster of stars (NGC 6530) is contained within the nebula. In the center is the 6th magnitude star 9 Sagittarii, which is thought to power the Lagoon Nebula, causing it to glow. It is approximately 6,500 light years from us and spans 60 light years across (Fig. 4.21).

Fig. 4.22 The dark dust lanes of the Trifid Nebula (M20) are clear in this image along with open cluster M21. The picture was taken in 1996 by the Burrell Schmidt telescope of Case Western Reserve University's Warner and Swasey Observatory located on Kitt Peak, near Tucson, AZ. (Image courtesy of REU Program/NOAO/AURA/NSF)

Only 1.5° north is the Trifid Nebula, designated M20 (NGC 6514). Much like the Lagoon Nebula, the Trifid is easy to locate, visible to the naked eye in dark skies, and is accentuated by a loose cluster of stars inside. This object is really two nebulous clouds joined together. The larger cloud looks intriguing, with its four dark dust lanes (although most observers see three) that stop just short of dissecting the cloud. In the case of both the Trifid and the previously mentioned the Lagoon nebulae, an O-III filter will enhance your view of the dust lanes. Not to be missed, just 0.5° northeast of the Trifid is a pretty little open cluster (M21). Although perhaps not as spectacular as the open clusters featured earlier, considered in the context of this star-rich region, it becomes an excellent cluster to observe (Fig. 4.22).

Before leaving this vast region rich in both nebulae and star clusters, we must realize that there are several dozen deep sky treasures waiting to be found. In an area so dense with activity, it can be refreshing to put down your map, move from star to star, and just observe. One suggestion is to move through the area that surrounds each star, in the teapot asterism of Sagittarius, sighting each area first in your binoculars, then in your telescope. Another method you may enjoy is to make slow, meaningful sweeps with your telescope at low magnification, in columns that run from south to north, with an occasional pause to observe with higher magnification if you so desire. The great thing about visual observing is that you are the pilot, and only you decide where you will go and how you will get there!

Object	Type	Right ascension	Declination
M54	Globular cluster	$18^h 55^m 03.3^s$	$-30° 28' 42''$
M70	Globular cluster	$18^h 43^m 12.7^s$	$-32° 17' 31''$
NGC 6652	Globular cluster	$18^h 35^m 45.7^s$	$-32° 59' 25''$
M69	Globular cluster	$18^h 31^m 23.2^s$	$-32° 20' 53''$
NGC 6563	Planetary nebula	$18^h 12^m 02.5^s$	$-33° 52' 06''$
NGC 6569	Globular cluster	$18^h 13^m 38.9^s$	$-31° 49' 35''$
NGC 6558	Globular cluster	$18^h 10^m 17.6^s$	$-31° 45' 47''$
NGC 6528	Globular cluster	$18^h 04^m 49.6^s$	$-30° 03' 21''$
NGC 6522	Globular cluster	$18^h 03^m 35.0^s$	$-30° 02' 02''$
M23	Open cluster	$17^h 56^m 56.0^s$	$-19° 01' 00''$
M24	Sagittarius star cloud	$18^h 17^m 00.0^s$	$-18° 36' 00''$
M25	Open cluster	$18^h 31^m 45.0^s$	$-19° 07' 00''$
NGC 6629	Planetary nebula	$18^h 25^m 42.4^s$	$-23° 12' 10''$
M22	Globular cluster	$18^h 36^m 24.2^s$	$-23° 54' 12''$

The coordinates shown are Epoch 2000.0.

In just this constellation of Sagittarius alone, there are 128 NGC and IC objects to be found, so be sure to consult your star atlas to find all of them. If you prefer a more organized approach, here is a short list of additional objects to search for in the Sagittarius constellation. They are listed in a way that you might observe if you start at the magnitude 2.6 star Ascella (Zeta [ζ] Sagittarii) and then circle the Sagittarius Teapot in a counter clockwise direction.

You may find some of these targets rather challenging to locate, but this will provide good practice in preparation for the deep sky objects featured in the next chapter—galaxies. Galaxies are usually much fainter and smaller than clusters or nebulae. So it is a good idea to spend some time locating and observing the objects in this chapter before trying to locate galaxies. Observing objects such as clusters and nebulae will build confidence in your ability to successfully locate challenging targets. Then if you target a galaxy and do not see it at first, you will know it is not necessarily because you lack the skill to find it, but rather it may be because of poor seeing conditions or insufficient dark eye adaptation. In other cases, it may simply be that a large aperture telescope is required to order to spot it.

For our next hunt, we will be searching for a planetary nebula. We will return to the constellation of Gemini, where earlier we used two of its stars (Castor and Pollux) to find the Beehive Cluster. The planetary we seek is the Eskimo Nebula (NGC 2392—see Fig. 4.1b), and is visible in the early spring when Gemini is well placed for viewing high overhead in the evening. One of the best ways to locate the Eskimo Nebula is to begin by sighting on magnitude 1.0 Pollux in Gemini. From Pollux go southwest about 8° until you arrive at the 3.5 magnitude star named Wasat (Delta [δ] Geminorum). Follow both of those steps first in binoculars, and then with your red dot finder or finderscope.

For the next steps be sure you are looking through the eyepiece at low power. Look for a triangle of stars less than 2° southeast of Wasat. The three stars that form the triangle are 63 Geminorum, 56 Geminorum, and 61 Geminorum. The Eskimo Nebula sits just outside the eastern leg of this triangle. After finding Wasat, move your telescope just 1.5° east to 5th magnitude 63 Geminorum. NGC 2392 is only 0.5° to the southeast and is well worth the challenge. In photographs, this magnitude 9.0 planetary nebula looks like a face surrounded by a fur parka, although in modest-size amateur telescopes this feature is much more difficult to distinguish.

Excellent observing conditions are necessary to see great detail in this planetary nebula. Although some like to use high magnification on this one, for optimum views medium power is recommended. If you increase the power too much you may find the view to be less than sharp. This is likely the most challenging object we have covered thus far, so do not get discouraged if you do not see it right away. As with other deep sky targets, you may need to review and follow each step again carefully in order to locate this one. Sometimes when you are sure that you are in the right spot but see nothing, a gentle wiggling or tapping of the telescope will bring an object to your attention.

Our brief tour of planetary nebulae would be incomplete without covering the awesome sight of one of the most photographed objects—the Ring Nebula, **M57**. In a 4-in. (102 mm) telescope the ring shape of this planetary is easily resolved. Although it also is magnitude 9.0, because it is larger than the Eskimo Nebula it appears brighter with more contrast and sharpness. In order to find this nebula, return to the Summer Triangle and put the star Vega into the field of view of the targeting finder on your telescope. Move to the east of Vega, where there is a group of four stars that form an elongated diamond shape. The star on the far eastern corner is Gamma (γ) Lyrae. Center this star in a low power eyepiece. The Ring Nebula is about 1° north—northwest, just before you arrive at Beta (β) Lyrae.

The Ring is bright, a little oval-shaped, and well defined. Using a 6-in. 'scope (152 mm) or larger may also reveal slightly lighter shades in the outer portions of the oval. Try using high power of 200× or more for a more stunning look at this object. If observing with a large telescope, another feature to look for is the nebu-

Fig. 4.23 The Ring Nebula in Lyra (M57). The magnitude 15 central star is a very challenging target for even very large amateur telescopes. (Image by the author)

la's central star (magnitude 14.7). You will need to use very high magnification and have great seeing conditions under a very transparent sky. It poses quite a challenge to be seen with large telescopes, although some have reported seeing it with a 14-in. 'scope. Even if the star will not reveal itself, you may find yourself returning to this nebula again and again before your session is over. The Ring Nebula is an object you will not want to miss. It will leave you in awe of the fact that this is only the beginning of many more exciting trips through the galaxy (Fig. 4.23).

There are many other superb nebulae scattered throughout our galaxy. After tackling the ones covered here, why not try for a few more in different areas? Some suggested targets include: NGC 7009 (Saturn Nebula, see Fig. 4.24), M16 (Eagle Nebula), M17 (Omega Nebula), IC 1805 (Heart Nebula) and IC 2118 (Witch Head Nebula).

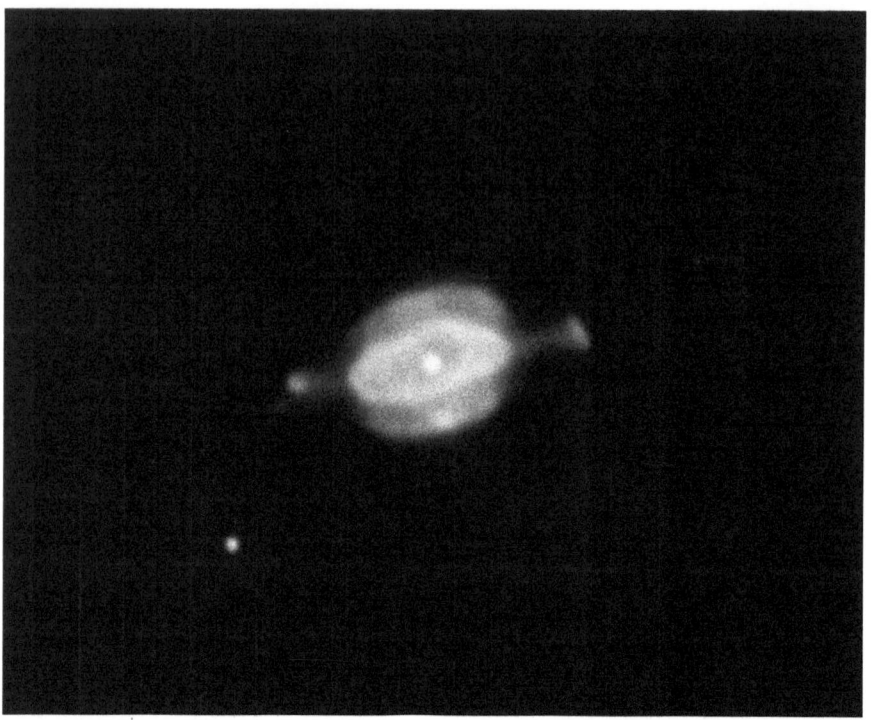

Fig. 4.24 The Saturn Nebula (NGC 7009) in Aquarius. (Image courtesy of NOAO/AURA/ NSF)

Helpful Things to Consider

Which stars will provide me with the easiest path to locate the nebula? Will I be searching for a large area of nebulosity or a small object, such as a planetary nebula?

After finding each target, did I record my experience and make notes so that it will be easier to find during my next observing session? Which nebulae would be the best ones to share with someone who is looking through a telescope for the first time?

As with star clusters, if you do not locate your target at first, retrace your steps and make a careful sweep of the area again. Remember that many nebulae can be illusive and require very dark skies, and sometimes a filter to be able to identify them. Be patient with yourself until you are finally able to see the object. Success is within your reach. Remember to take notice of any extra celestial gems you may observe in the process.

All of the deep sky targets in this chapter can be found in small- to medium-sized telescopes, but of course the larger the aperture, the more breathtaking the view. For an interesting comparison, try pairing up with a friend who has either a larger aperture or a different type of telescope than you do. The best way to enjoy views through various telescopes is to visit your local astronomy group. Most clubs have websites that are updated on a regular basis with current times and locations of meetings, upcoming star parties, and helpful tips on backyard astronomy. There can be a very high excitement level at star parties when so many people are sharing views of their favorite objects through their telescopes. Within the club you are sure to find people that are very helpful to newcomers, and more than happy to share their ideas, suggestions, and experiences.

Chapter 5

Whirlpools and Sombreros: Searching for Galaxies

It was over 1,000 years ago, during the tenth century AD, that we find the first written observation of the nebula that we now refer to as the Andromeda Galaxy (M31). The astronomer was Al Sufi of Persia, and he called it the "Little Cloud." Centuries later, on October 13, 1773, Charles Messier discovered the Whirlpool Galaxy (M51). His observing partner and associate Pierre Méchain, discovered the Sombrero Galaxy (M104) on May 11, 1781. In the mid-nineteenth century William Parsons (Lord Rosse) built a 1.8-m telescope, the world's largest at the time. After observing the Whirlpool Galaxy's distinctive spiral structure in 1845, he published his famous sketch of M51 in 1850, which includes both the larger galaxy NGC 5194 and its smaller companion NGC 5195. Although William Herschel cataloged hundreds of galaxies, neither he nor his contemporaries knew their true nature (Figs. 5.1 and 5.2).

In 1919 Harlow Shapley of the Mount Wilson Observatory published a study that indicated a much larger size for our Milky Way Galaxy than previously thought. The result of his measurements of Cepheid stars in dozens of globular clusters led him to conclude that our galaxy is about 326,000 light years in diameter. Today we know that it has a diameter closer to 100,000 light years (except for a few straggling stars much farther out). However it was perhaps this revelation of a far larger Milky Way Galaxy that caused him to believe spiral nebulae were inside of our own galaxy and not independent star systems.

In 1920 Heber Curtis of the Lick Observatory took the opposite view of Shapley, and the two engaged in what has been referred to in modern astronomical history as the Great Debate, held at the National Academy of Sciences in Washington, D.C. Curtis promoted the idea that the spiral nebulae were not a part of our Milky Way, and that any lack of these spiral nebulae in the area of the Milky Way was due to obscuring dust hiding them from our view. He had observed these dust lanes in

© Springer International Publishing Switzerland 2015
D. A. Jenkins, *First Light and Beyond*, The Patrick Moore Practical Astronomy Series,
DOI 10.1007/978-3-319-18851-5_5

Fig. 5.1 The Whirlpool Galaxy M51 in Canes Venatici. The larger galaxy is NGC 5194, while the smaller one is NGC 5195. (Image courtesy of NOAO/AURA/NSF)

other spiral nebulae (galaxies) and so concluded that our own galaxy must have the same or a similar component.

Astronomers did not have to wait very long before the debate was settled. Clarity arrived when Edwin Hubble began his own examination of Cepheid variables in early 1924. With this new data he calculated that the Andromeda nebula (now known as the Andromeda Galaxy, M31) must be at least 1 million light years distant and

Fig. 5.2 The Sombrero Galaxy—M104—lies about 50 million light years away in the Virgo constellation. (Image courtesy of ESO/P. Barthel. Acknowledgments: Mark Neeser [Kapteyn Institute, Groningen] and Richard Hook [ST/ECF, Garching, Germany])

therefore outside of our own Milky Way Galaxy. That year, like many other pivotal ones that came before it, would forever change astronomy and how we view our place in the universe. Edwin Hubble went on to classify galaxies into a few main types—elliptical, spiral, barred spiral, and irregular. Within those general classifications are many different variations. During the last century astronomers have observed a vast number of galaxies, and most of them do not seem to fit what may come to the average person's mind if asked to describe a galaxy.

In 1959 Russian astronomer B.A. Vorontsov-Velyaminov (1904–1994) cataloged 355 galaxies that actually interact. Later that catalog grew to 852 such galaxies, and then to 2,014 (posthumously). Vorontsov-Velyaminov is also well known for the immense work *Morphological Catalogue of Galaxies (MCG)*, containing 34,000 galaxies of magnitude 15.0 or brighter—well within reach of very large amateur telescopes. In 1966 famous astronomer Halton C. Arp (1927–2013) published his *Atlas of Peculiar Galaxies*. This list of galaxies calls attention to some of the most unique and peculiar-looking shapes of galaxies that can be observed in amateur telescopes. The *Deep Sky Field Guide to Uranometria 2000.0* often refers to galaxies from these catalogs along with several others. In total, this field guide lists (and also plots in its companion *Uranometria 2000.0*) 25,895 galaxies, enough to keep any amateur astronomer busy for many years.

Like nebulae, galaxies are a unique and mesmerizing type of deep sky object that you will enjoy searching for and observing. They twist and stretch. They dance and merge. Equipped with a telescope and with no shortage of cataloged galaxies, we have a front row seat to the show.

We will begin our quest for galaxies with some of the easiest to find, and progressively move towards those that are more challenging to locate and observe. First, however, let's consider a few tips to help you get the most from the view when you observe.

Rely on a Good Star Atlas

In the previous chapter we covered a fair amount of nebulae and star clusters. If you have been able to do some observing of your own, you may have come across a galaxy or two, knowingly or unknowingly, while observing a star cluster or nebula. For example, in our discussion of globular cluster M13 in Hercules, we mentioned the nearby spiral galaxy NGC 6207. Although it is hard to miss in Fig. 5.3, we are reminded of the need to be aware of everything in our field of view, because we never know what surprises lurk nearby. It is easy to miss something like NGC 6207 in the eyepiece if you are not aware of it or are not looking carefully. For those reasons a good star atlas or astronomy software is essential, so that you can know what to expect in a given area, identify unknown objects that you come across, and have plenty of options for new targets to observe while in the field.

Modern star maps, whether printed in an atlas or displayed within software, usually do a pretty good job of illustrating the size of a galaxy to match the scale of the map itself. Larger symbols denoting galaxies can be expected to represent those that will be easier to spot purely by size. This is not to say that the galaxy will be bright. Remember, its listed magnitude may or may not be bright at all, and if it has a low surface brightness, it will be a little more challenging. Galaxies illustrated with small representations should be considered quite illusive. In Fig. 5.4, compare the size of M31 with M110 or M33. In this star chart, the galaxies are drawn to scale. The Andromeda Galaxy actually is about 3.5° × 1°, whereas M33 is smaller, at about a size of 1° × ½°. Spend a few minutes getting to know this and the other features in your own star atlas or software-generated star charts. This will help you to become adept at using the charts at the telescope and to have a more realistic expectation of the characteristics of the galaxy you are searching for.

Fig. 5.3 The galaxy NGC 6207 is clearly visible in this picture along with globular cluster M13 in Hercules. (Image by the author)

As we have learned already, it is essential to learn to recognize star patterns as you move from one field of view to the next. This is true whether that field of view is large, as in the case of binoculars, or a much smaller view through an eyepiece. Although star hopping from one individual star to another is at times necessary, you will find your targets faster and be more sure of their location if you try to match the patterns of several stars that you see on the chart with what you see in your binoculars, finderscope, or eyepiece.

Developing star pattern recognition makes star pattern jumping much easier, especially if it does turn into a little bit of star hopping. If this was an important skill to develop for locating star clusters and nebulae, it is equally if not more important in order to be successful at finding galaxies. These are faint objects, and

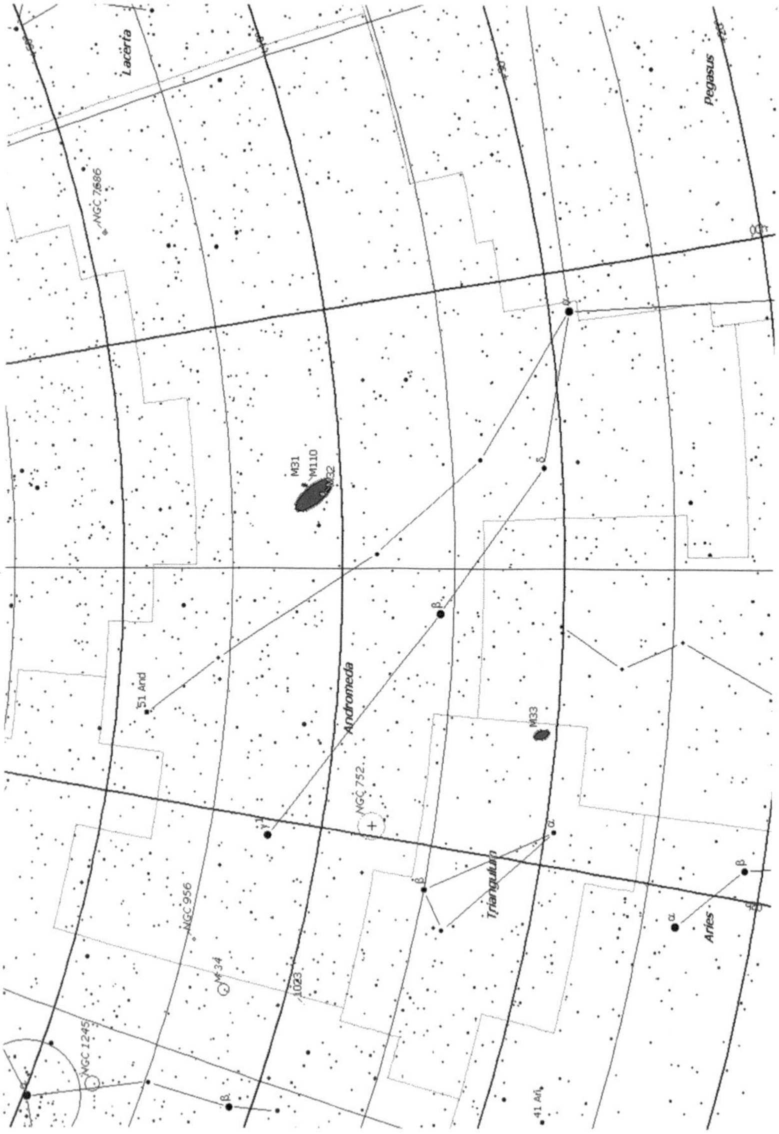

Fig. 5.4 Note how the star chart indicates the comparative size of each galaxy according to scale. (Star chart provided courtesy of *Sky Tools 3* at www. skyhound.com)

most of the time will not wave at you like the E.T. Cluster (NGC 457). Like some of the planetary nebulae you may have observed from Chap. 4, many galaxies have a low surface brightness, which makes them harder to recognize at first. With patience and persistence, you will see them. When you do, prepare to be amazed, as many of them will leave you in awe of our ability to see these galactic islands with backyard instruments.

Here are a few additional tips, and some mentioned earlier that are worth repeating:

- If at first you do not see a galaxy, use averted vision. Even if you readily spot the galaxy, use averted vision—it may reveal new details that you did not see initially.
- Avoid fatigue. Sit down if possible while observing, or at least take breaks from standing as needed.
- If eye strain is a problem for you, try an eye patch to prevent strain on the unused eye.
- If you have trouble spotting a galaxy, take a little stellar excursion around the area nearby and then return to the location after a few minutes. Sometimes a little detour and a fresh look at your target area can help make the galaxy stand out.
- A little tap or jiggle of your telescope can help reveal an object hiding in plain sight.
- Take your time observing the object—details in the eye accumulate over a period of several minutes.
- Breathe deeply so that you take in plenty of oxygen and remain calm.
- Sketching the galaxy will help you to see more detail.
- Examine the galaxy under different magnifications. Using a set of eyepieces that are parfocal will allow you to do this with the least amount of disruption to your view. Eyepieces that are described as being parfocal are designed so that you can change out any eyepiece in the group without the need to re-focus the image.
- Keep in mind that published magnitude numbers can be subjective, and so some objects that are indicated to be too faint for your telescope may in fact be visible; other objects that should be within range of your equipment may not be visible because of various factors such as poor sky conditions, light pollution, experience, etc.
- Be sure that any nearby light sources are extinguished or blocked from contaminating your view.
- If you will be doing deep sky observing, protect your eyes from sunlight that day or even a day or two prior to observing, so as to make your night vision as sharp as possible.
- The initial aiming process first involves the naked eye. Look at your map, and then look at the sky and see how much you recognize without optical aid. Then re-examine the area with binoculars (greater mobility, comfort, and ease of use). Confirm what you see in binoculars by viewing the same area in your finderscope. Finally, enjoy the view through the eyepiece.

A good standard observing routine of the object itself should begin with viewing an object with direct vision, then averted vision, and back to direct vision again. Next, step away from the eyepiece for a few moments, then repeat those steps. It is also very important to make notes or recordings of your observations. Many details can be forgotten if you rely on memory alone. You may be surprised at the amount of detail that you saw after you review your notes or a recording much later.

After gaining some observing experience, it's enlightening to go back and review your notes. Observers often wonder how details missed at one time seem so obvious months or years later.

Now let's begin to explore several of these exciting star cities. If you own a telescope that has setting circles, you can simply use the coordinates as found in a star atlas or astronomy software program, and turn your 'scope directly to the object. A GOTO 'scope has the coordinates of many objects stored in its memory. However, it is not only educational to try and find the objects with just a star chart and the aid of this book, but it is sure to be a more fun experience as well. Some say life is about the journey, and in many respects so is astronomy. There is a certain thrill that comes from hunting down an object for yourself, being able to look through the eyepiece, and say "I found that!" This leads to a euphoric sense of fulfilment that is enduring.

Some of the Easy Galaxies

Our first galaxy to review is the closest naked-eye galaxy in our night sky, **M31** (NGC 224). The Andromeda Galaxy is just 2.5 million light years away, shining brilliantly at magnitude 3.4. It, along with M33 and our own Milky Way, are the largest galaxies that are a part of what is known as the Local Group of approximately 50 known galaxies (many of them are satellites of the three largest), all within a 3-million-light-year diameter. It appears that we are on a collision course with the Andromeda system, thought to take place in about 4–5 billion years, although individual stars are so far apart that stellar impacts would rarely occur.

Because Andromeda is so close to us, astronomers have been able to observe many of its star fields and measure the distance of stars within it. This in turn provides a measuring stick to obtain distances to other galaxies. There are 509 globular clusters that have been detected within the Andromeda Galaxy, and several can be seen in large amateur telescopes of 10 in. (254 mm) or more. One example is the globular cluster Mayall II, which shines at magnitude 13.7. More details on this topic can be found in the book *The Andromeda Galaxy* by Paul Hodge, the article by Alan Whitman entitled "Exploring Messier 31" published on p. 59 of the November 2013 issue of *Sky & Telescope*, and also in the 1985 study on the globulars of M31 (Crampton et al. [6]) (Fig. 5.5).

The Hubble Telescope has revealed that the Andromeda system is a staggering 150 million light years across and contains more than 400 billion stars. At the galaxy's core is a supermassive black hole surrounded by a large glowing blue light, which in 2005 was determined to be a mass of more than 400 hot blue stars moving at a rate of 2.2 million miles (3.5 million km) per hour.

Fig. 5.5 Our sister galaxy M31 in Andromeda, visible here with its two companion galaxies. (Image by the author)

Locating this large phenomenon is easy. Find the star Alpha (α) Pegasi within the Great Square of Pegasus. Then cast your gaze 14° northeast until you see the unmistakable glow of M31. Even in binoculars or a small telescope the view is excellent, and through the eyepiece of a 6-in. (152 mm) or larger 'scope the view is breathtaking. It fills up the field of view in an eyepiece so much that it is difficult to see it all at one time unless you are using an extremely wide field eyepiece. As you pan from one side of the galaxy to the other, note its companion galaxies, the smaller M110 (NGC 205) and the much smaller M32 (NGC 221). Under dark skies and excellent observing conditions, dark lanes in this galaxy should be visible. This is definitely number one on the list of easy galaxies to observe in autumn and winter.

The next galaxy on our list is also one that can be seen with the naked eye, **M33** (NGC 598), sometimes called the Pinwheel Galaxy) in the Triangulum constellation, 2.7 million light years away. It shines at magnitude 5.7 and so is considerably fainter than the Andromeda Galaxy, but is visible to the eye under dark skies. It is a beautiful spiral galaxy, although just 50 million light years across. Because it is also so close to us, astronomers are able to gather a lot of information about its stars. Most interestingly, recent studies show that much like our own Milky Way Galaxy, M33 has lots of star-forming regions, many planetary nebulae, and quite a few emission nebulae that are supernova remnants. It, too, has several objects within it that can be observed with large amateur telescopes, and many of these features are NGC or IC catalog objects.

Fig. 5.6 Another galaxy in our Local Group – M33 in Triangulum. (Image by the author)

Viewing any of these fainter secondary features in this galaxy requires a 12-in. (305 mm) 'scope or larger. Among many examples are NGC 595, NGC 592, IC 135, IC 143, and NGC 604. NGC 604 is a very active, very large region of star birth inside M33 that is about 1,500 light years across (the Orion Nebula is only 13 light years across). There are at least 200 hot blue stars inside of it, and astronomers estimate the age of those stars to be about 4 million years old. There is also a population of red supergiant stars inside that are more than 12 million years old.

M33 is easy to find because of its proximity to the Andromeda Galaxy. Use the chart in Fig. 5.4 to help. From M31 go about 15° southeast, and with binoculars you should pick it up easily or with the eye alone using averted vision. The view of M33 in a small telescope is not as dramatic a sight as M31 because of its lower surface brightness, so to really appreciate this spiral galaxy, try viewing it through a larger telescope of 8 in. (203 mm) or more (Fig. 5.6).

Now we turn to the constellation of Hydra, a galaxy visible in the spring and summer, located 15 million light years away. Although originally discovered by French astronomer Nicolas-Louis de Lacaille in 1752, Charles Messier observed this galaxy on February 17, 1781, and it became the 83rd entry in his catalog. The size of **M83** (NGC 5236) spans a distance of 40,000 light years, smaller than our own galaxy but an impressive one to observe. There have been at least six supernovae observed in M83 since 1923. It is sometimes called the Southern Pinwheel Galaxy, shining at a magnitude of 7.5. Some experienced observers have reported being able to see this galaxy with the naked eye. This of course is only

Fig. 5.7 Galaxy M83 in Hydra. (Image courtesy of ESO/IDA/Danish 1.5 m/R. Gendler, S. Guisard [www.eso.org/~sguisard] and C. Thöne)

possible for the well trained eye, under the darkest skies and excellent observing conditions. If you are observing from a location very far north, this galaxy may be subject to poor sky conditions near the horizon because it is located low in the southern sky. Otherwise, binoculars should make this sharp looking galaxy stand out. Clearly visible in small telescopes, and in a 6-in. or larger the spiral structure is prominent. It also responds well to medium and high magnification (Fig. 5.7).

Up until this point we have not offered the same amount of detail for locating galaxies as we did for clusters and nebulae, partially because the first two (M31 and M33) galaxies are fairly easy naked-eye objects. A second reason is that it allows you to practice finding these targets with your own star atlas or computer-generated star charts, in addition to the charts sometimes included here. For those reasons we'll continue to do this occasionally so that you can build your skill in star pattern recognition and star pattern jumping with the use of your own charts. However, let's now explore how to find our current target with a little more detail. M83 is located 18° due south of the prominent magnitude 1.0 star Spica in Virgo, and is also 15°

southeast of Beta (β) Corvi. Let's narrow in on our target and see what patterns emerge in the surrounding area within 10° of M83. From β Corvi move 10° due east to Gamma (γ) Hydrae, a magnitude 3.0 star.

Just 4° southeast is a group of stars forming a pattern that looks like the number 7. The stars that make up this pattern all fit within about a 2° diameter. From the bottom of the "7" pattern, M83 sits just 1.5° southeast, framed between two stars that are approximately sixth magnitude. Take a look at Fig. 5.8a, b for a closer look. The first chart is a broader view of the area, while the other is a close up of the star field surrounding M83.

In both charts the numbers in parenthesis and also any numbers appearing after Flamsteed designations are the magnitudes of the stars they are beside (be sure to move the decimal over one place to the left). Once you have explored the area in binoculars, aim your finderscope on the stars that form the "7" and continue southeast until the center mark of the finder is midway between the two 6th magnitude stars mentioned earlier that form a frame around M83. An eyepiece with a 1° field of view will easily contain both M83 and the sixth magnitude star on its eastern side.

So far we have discussed three galaxies, two of which are easy to see with the naked eye, M31 at 2.5 million light years away and M33, which is 2.7 million light years away. Although it may be possible to see the magnitude 7.5 galaxy M83 with the eye alone, it is difficult for most observers who may not have regular access to extremely dark skies and who may also not be far south enough for the object to be clear of hazy conditions. If you do have the right conditions to view it, count yourself fortunate to be among the few who have seen an object 15 million light years away with just the naked eye! However, for most observers in mid-northern latitudes, there is a more attainable candidate for the most distant naked-eye galaxy. According to new data from Hubble images in 2007, it is 11.6 million light years away. It was discovered in 1774 by German astronomer Johann Bode and is commonly called **Bode's Galaxy**, or **M81** (NGC 3031). At magnitude 6.9 it has also been observed by several experienced observers with the eye alone.

One way to test whether or not you are really detecting an object at the threshold of naked-eye visibility is to see if you can point your finderscope at what you believe to be visible. In other words, if you think you are detecting M81 with your eyes, see if you can target the crosshairs of your finderscope right on that spot in the sky. You should then be able to see it within a ½° of the view within the eyepiece. A similar comparison can be done with binoculars, by toggling back and forth from the naked-eye view to binoculars. In this way you can determine if your eyes still see the galaxy when the binoculars are lowered. The only thing left to do is to be sure that you are not willing it into view simply because you know it should be there. In that regard, another test of your objectivity would be to observe on a night of poor transparency. On that night your eyes should not be able to detect it even though you may still know where it is based on your knowledge and memory of its proximity to nearby stars (Fig. 5.9).

In recent years, Bode's Galaxy has continued to attract attention for several reasons. First, the nearby **M82** (NGC 3034, or the Cigar Galaxy) has had four recorded supernovae thus far, the most recent being the supernova explosion SN 2014J.

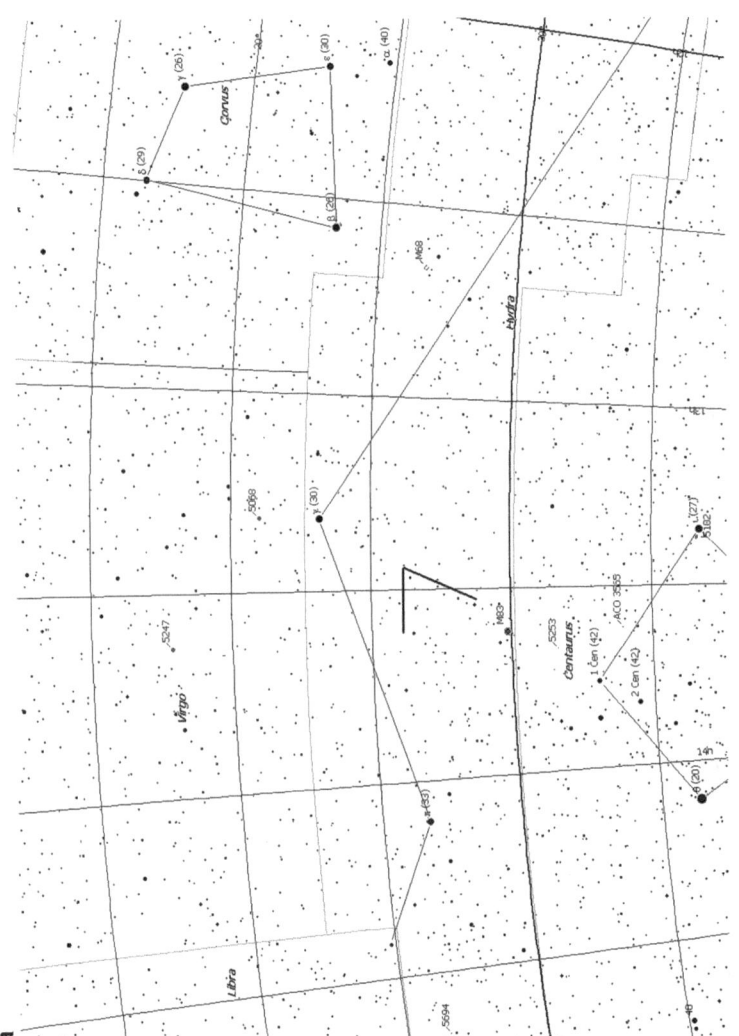

Fig. 5.8 (a) The Hydra constellation, indicating a pattern of stars leading to the galaxy M83. (Star chart courtesy of *Sky Tools 3* at www.skyhound.com.) (b) Enlarged area near M83. (Star chart courtesy of *Sky Tools 3* at www.skyhound.com)

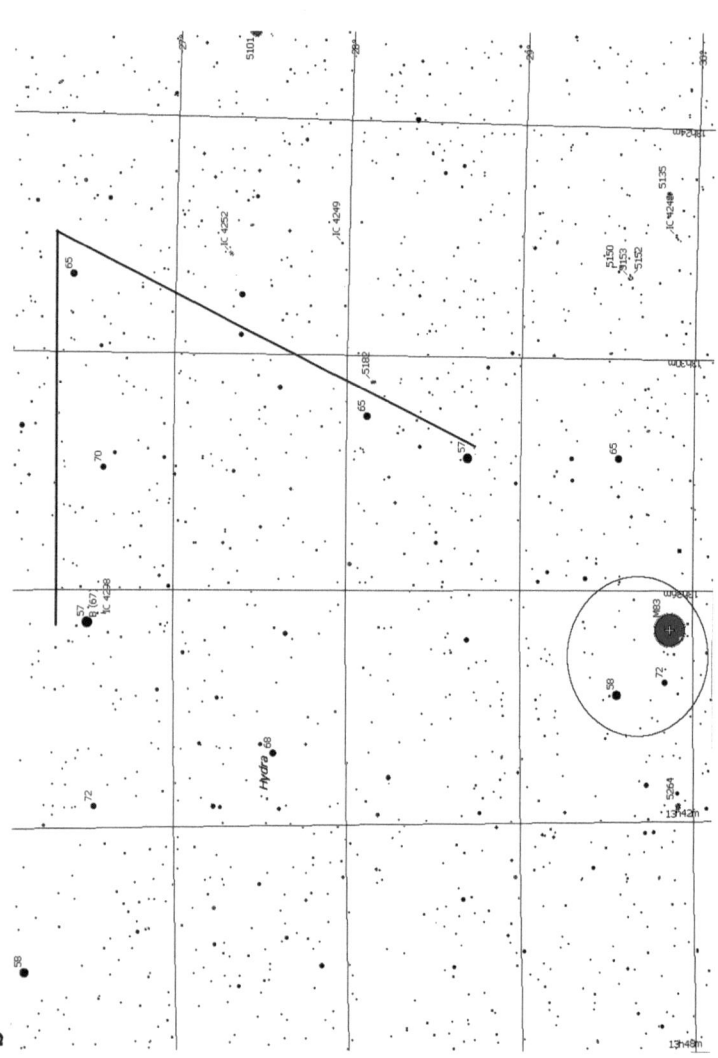

Fig. 5.8 (continued)

Fig. 5.9 Galaxy M81 in Ursa Major. (Image courtesy of NOAO/AURA/NSF)

By February of 2014, it brightened to an apparent magnitude of 10.5. This is one of the closest supernovae to be visible in decades and will provide data for astronomers to use in refining our ability to measure distances and better understand the processes involved in these stellar explosions.

Another reason that makes M81 such an interesting galaxy is that it is related to both M82 and NGC 3077. Both galaxies are fainter; M82 is about magnitude 8.5 and NGC 3077 is about magnitude 10, but you should be able to see both of them along with M81 in a very wide field eyepiece. Increase the magnification to bring out more details in each one. Research by astronomers has found that about 200–300 million years ago M81, M82, and NGC 3077 had a close encounter, which has left a trail of stars strewn about in an area called Arp's Loop. In 2008 Hubble imaged the mysterious clumps of blue stars that seem to form a stellar bridge between the galactic partners. At least 14 dwarf galaxies are a part of the M81 Group; several of them are thought to be forming as a result of the close interactions of these three.

To locate M81 and its close companions, begin by finding the Big Dipper and find second magnitude Dubhe (Alpha [α] Ursae Majoris). About 10.5° west is the tenth magnitude star 23 Ursae Majoris. Just under 7° due north is the fifth magnitude star 24 Ursae Majoris. M81 lies 2° to the southeast of 24 Ursae Majoris. NGC 3077 is 40′ southeast of M81, and M82 is 30′ north of M81. Figure 5.10a, b show both a broad view and a close up map of the area. Several other fainter galaxies are also shown on the map. Once you are comfortable with the easier galaxies, push your observing skills a little and try for some of the fainter ones in this area, such as IC 2574 or NGC 2976.

Now let's turn our attention back to the Whirlpool Galaxy—**M51** (NGC 5194)— and its smaller companion **NGC 5195** (see Fig. 5.1). These two dynamic galaxies are in fact orbiting one another during periods that last for approximately 400 million years. The smaller NGC 5195 is about 400,000 light years further out than the larger NGC 5194, although the two are still connected by stellar tidal interactions. The most recent measurement for M51 places it at 27 million light years from us. The last supernova in NGC 5195 was in 1945, but NGC 5194 has had four supernovae. The most recent was in 2011 (SN 2011dh), probably caused by the explosion of a yellow supergiant star. According to a study in 2013 (Van Dyk et al. [7]), a yellow supergiant was captured by Hubble but is no longer visible. However, in 2014 a blue point source was detected in the ultraviolet wavelength and is thought to be the companion star to the yellow supergiant that exploded.

Observing these two interacting galaxies is a real treat. Small telescopes will easily reveal this magnitude 8.4 galaxy. To uncover lots of detail in this spiral, use a 10-in. (254 mm) telescope or larger. You will find M51 in the constellation Canes Venatici, which borders Ursa Major. Find the second magnitude star Alkaid (Eta [η] Ursae Majoris) in the ladle of the Big Dipper and then look west just 2° to the fifth magnitude star 24 Canum Venaticorum. At this point binoculars should show M51 just 2° south of that star. Use the charts in Fig. 5.11a, b to help you locate M51.

M101 (NGC 5457) is another galaxy you will not want to miss and is located not too far from M51. Please keep in mind that this one is much more challenging to find with smaller telescopes. Although its visual magnitude is 7.9, M101 (sometimes called the Pinwheel Galaxy) has a much fainter surface brightness than the Whirlpool Galaxy. However, what it lacks in surface brightness is made up in larger apparent size (28.8′ × 26.9′) and this is one reason that makes it such an interesting galaxy to observe. This one screams for the use of averted vision. With a lot of patience and persistence you will be able to see it in a 6-in. 'scope and try at least 10 in. of aperture to see the details in its beautiful spiral structure. The photo in Fig. 5.12a gives a good approximation to how it will likely appear in a medium-size telescope.

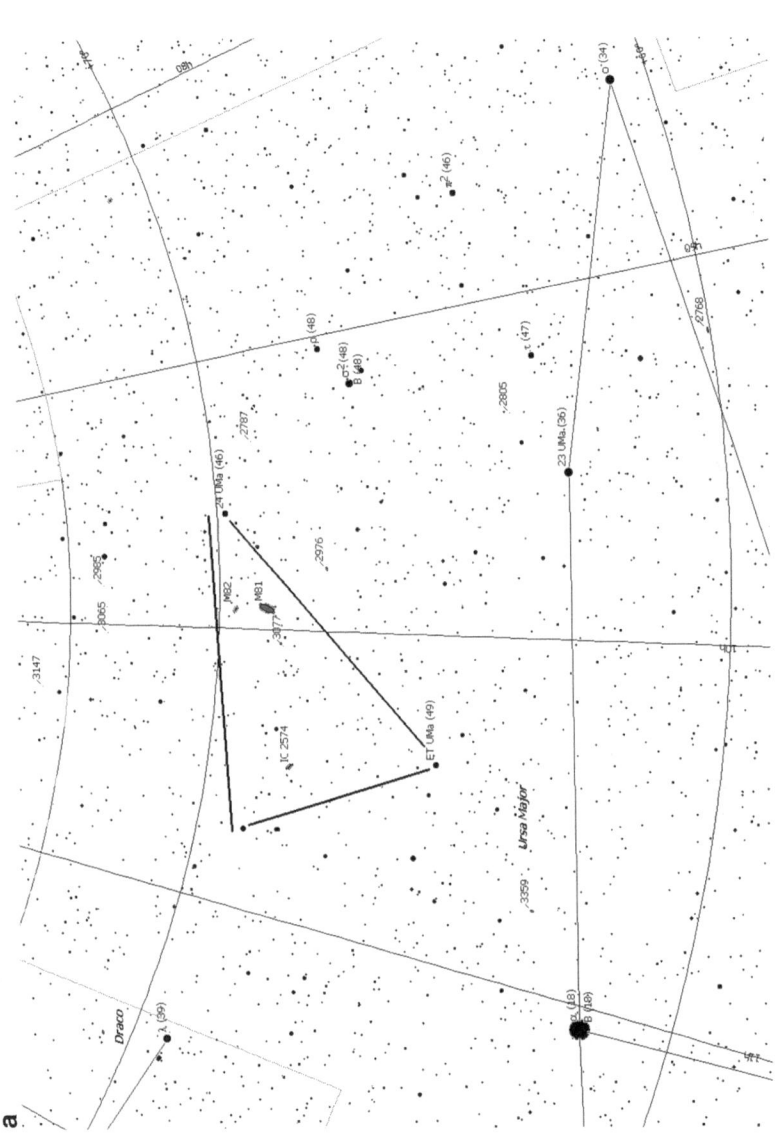

Fig. 5.10 (a) Broad view of the area near M81 in Ursa Major. (Star chart courtesy of *Sky Tools 3* at www.skyhound.com.) (b) Enlarged area near M81. (Star chart courtesy of *Sky Tools 3* at www.skyhound.com)

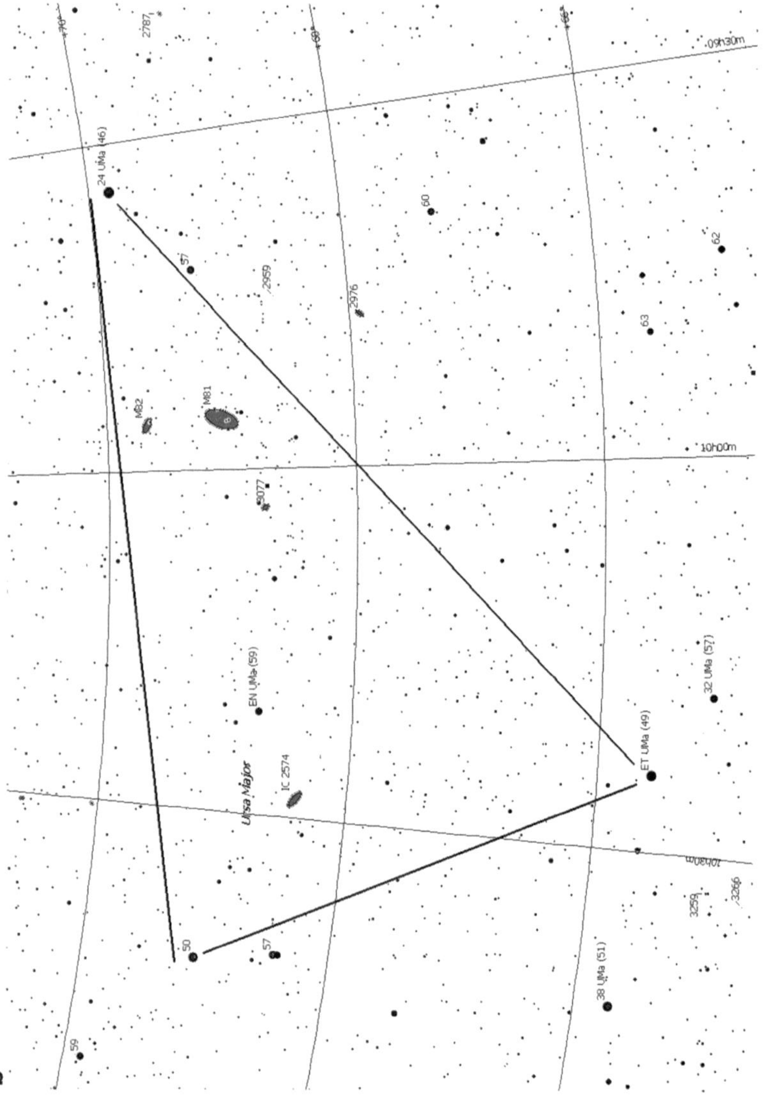

Fig. 5.10 (continued)

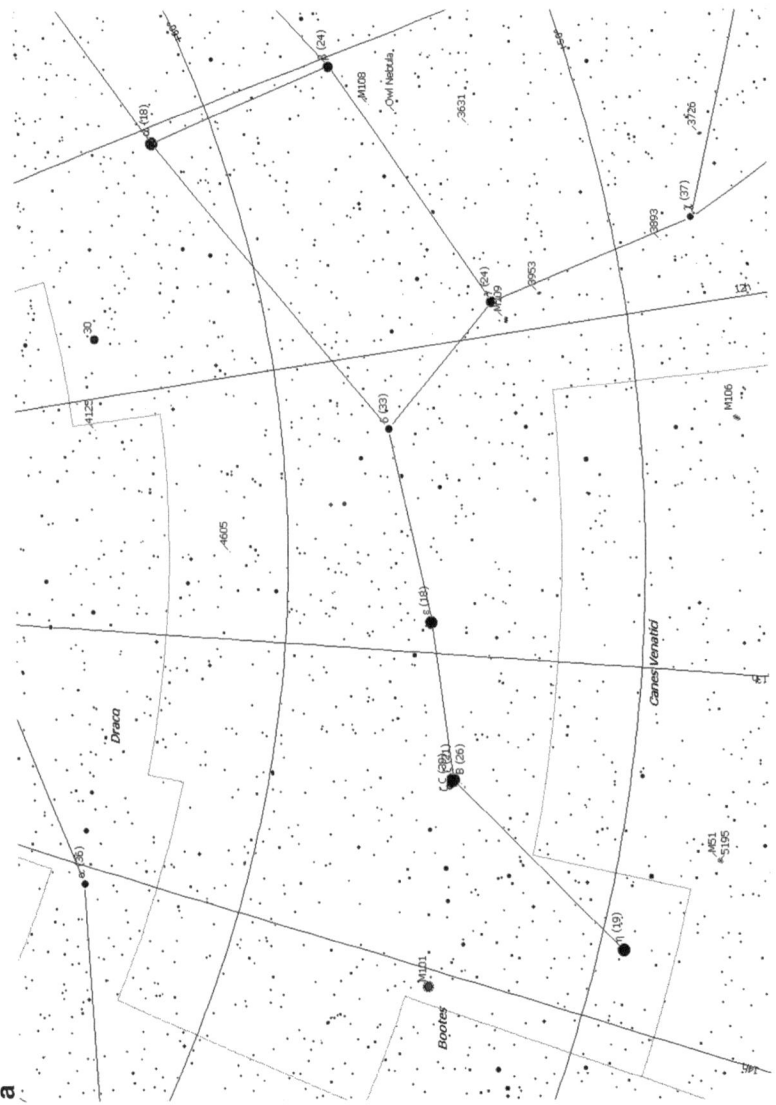

Fig. 5.11 (a) Broad view of the Big Dipper. M51 can be seen located near the star Eta (η) Ursae Majoris. (Star chart courtesy of *Sky Tools 3* at www.skyhound. com.) (b) Enlarged view of the area near M51 in Ursa Major. (Star chart courtesy of *Sky Tools 3* at www.skyhound.com)

b

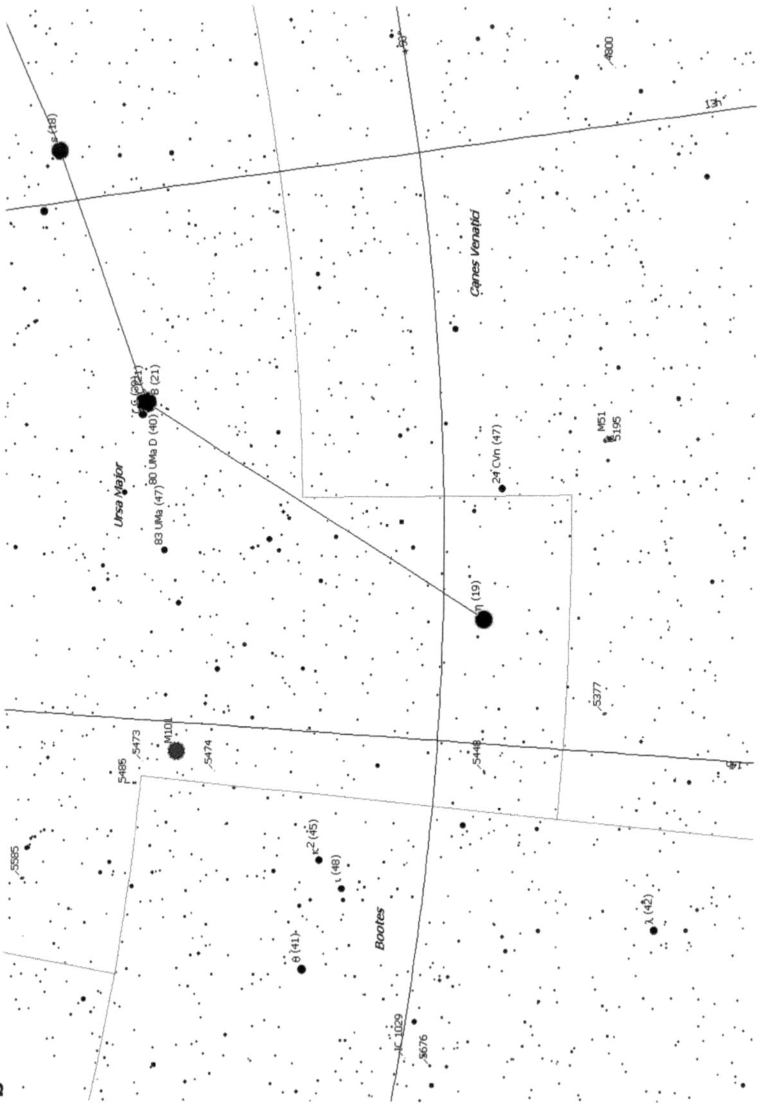

Fig. 5.11 (continued)

Fig. 5.12 (**a**) The large galaxy M101 in Ursa Major. (Image by the author.) (**b**) Enlarged area near M101. (Star chart courtesy of *Sky Tools 3* at www.sky-hound.com)

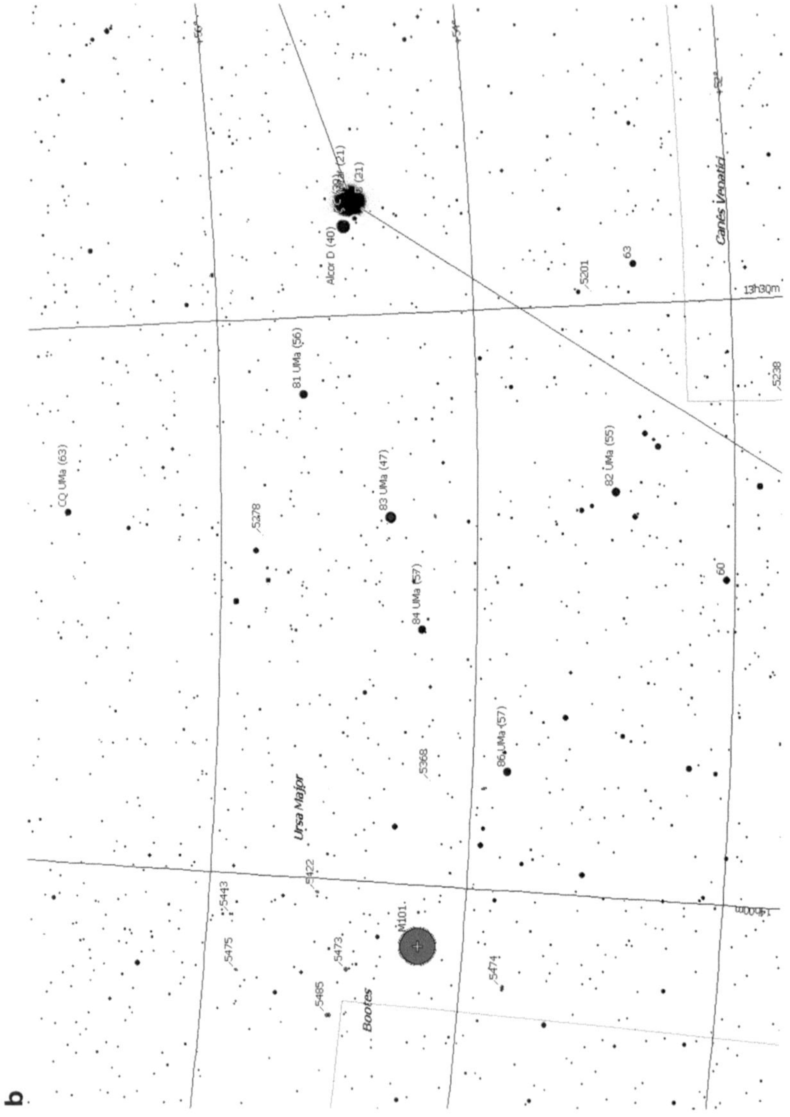

Fig. 5.12 (continued)

M101 is a large face-on spiral galaxy that extends over 184,000 light years in size. The examination of data from the fourth discovered supernova (SN 2011fe) in M101, has allowed confirmation of its distance, which although still not known precisely, is now estimated to be 22 million light years away. Our understanding of how galaxies grow and change over time is still being refined. The age of stars within certain areas of M101 presents a puzzling example that astronomers have also found in other galaxies. The central bulge of M101 is younger than the disk that surrounds it. This bulge contains lots of dust, an indicator of heavy new star formation. The inner disk outside of the bulge has an older population of stars (6–8 billion years), while the outer spiral arms contain much younger stars (2–4 billion years).

To find M101 begin with the broad chart in Fig. 5.11a. You will find it 5.5° east of the fourth magnitude Alcor and its double star companion, second magnitude Mizar. Between Alcor and M101 is a row of stars between magnitude 4.7 and 5.7 that cascade from northwest to southeast. M101 is exactly 1.5° northeast of the last star in the group of four—86 Ursae Majoris. Figure 5.12b shows a close up of this area.

Our next stop is estimated to be at least 28 million light years from Earth in the constellation of Virgo and is pictured in Fig. 5.2. **M104** (NGC 4594) is the Sombrero Galaxy, noted for its massive bulge and distinctive dust lane that we see tilted 6° above its equatorial plane. It was discovered on May 11, 1781, by Pierre Méchain. An estimated 2,000 globular clusters hover around this galaxy that extends 50,000 light years across, as it recedes from us at 1,000 km per second. Our galaxy only has about 200 known globular clusters. Inside the Sombrero Galaxy's center lies a black hole with the mass of one billion suns. After a view through the Spitzer Space Telescope's infrared vision in 2012, some astronomers have concluded that this galaxy is elliptical, with a large dusty disk embedded inside of it. The total mass of this galaxy is 800 billion suns.

The eighth magnitude Sombrero Galaxy is located 11° west of first magnitude Spica in Virgo. However it can be a little challenging to find from that direction. A better way is to start with the four stars in Corvus that resemble a kite leaning on its western side. With binoculars find third magnitude Algorab, the northernmost star in this group of four. Look just 3° north of Algorab to find a triangle of stars of about sixth magnitude that point to the northeast. Figure 5.13a, b should be of great help in finding this stunning galaxy.

The star 56 Corvi sits at the north apex of this triangle, itself pointing northeast to a tiny compact triad of three stars. M104 is just 1° northeast of this marker. From another standpoint, our target is also 1.5° north of VV Corvi, a fifth magnitude variable star. Once you find this object the first time, you will be able to return more easily because you will quickly recognize the three signposts that lead to it:

1. The third magnitude star Algorab in Corvus.
2. The 1° triangle formed by three 6th magnitude stars just north of Algorab.
3. Tiny, compact triad of stars just 1° northeast of the star 56 Corvi.

Visually, this galaxy is immediately recognizable with its prominent dust lane across the center in small to medium telescopes. However, for a stunning view of this feature, use a 10-in. or larger 'scope. At medium powers it presents a view that

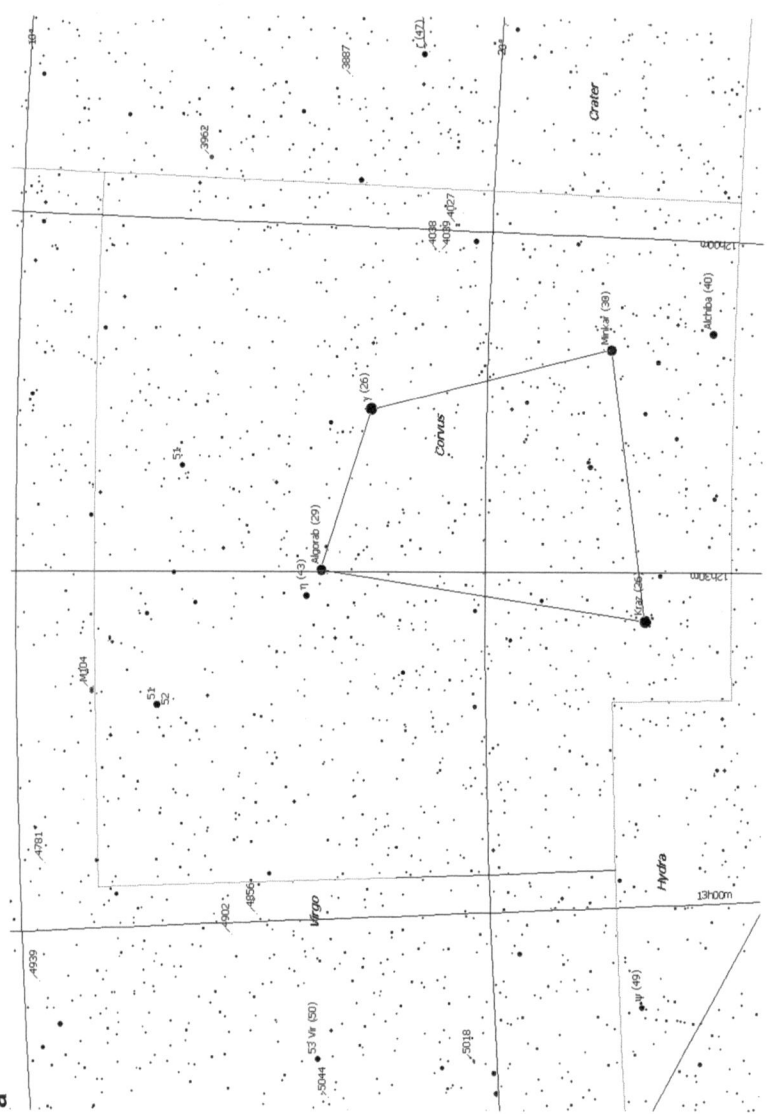

Fig. 5.13 (**a**) The Sombrero Galaxy is shown here just north of the star Algorab in the constellation of Corvus. (Star chart courtesy of *Sky Tools 3* at www. skyhound.com.) (**b**) Enlarged area showing the route to the Sombrero Galaxy M104. (Star chart courtesy of *Sky Tools 3* at www.skyhound.com)

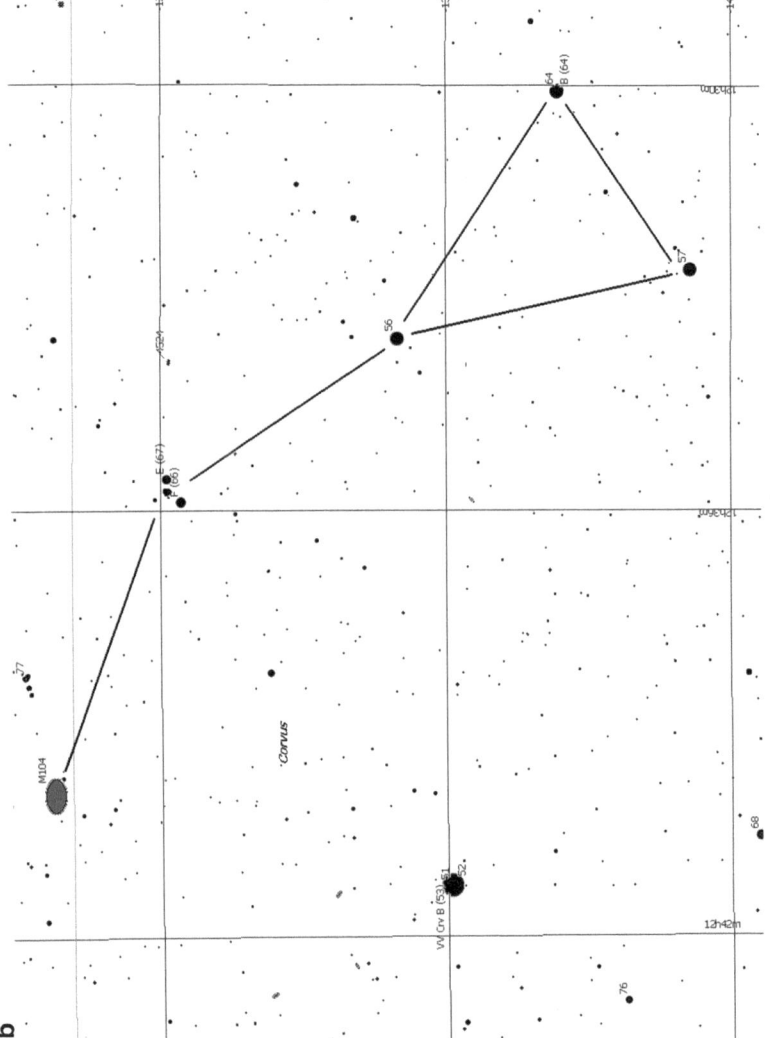

Fig. 5.13 (continued)

appears three dimensional, like a giant orb floating in a stellar ocean. This is the kind of object that tends to fire up enthusiasm for visual astronomy in anyone who sees it!

After having considered these eight galaxies in detail, it may be a good idea to reflect how you are comprehending the use of star pattern recognition. Glance back through some of the figures containing the objects that have interested you the most or that you'd like to try to find first on your next night of observing. Do you remember the suggested routes to take to each galaxy? What patterns stood out in your mind that will act as signposts to take you to your destination? Have you found different star patterns than the ones recommended so far? That's good—keep doing so, and feel free to use whatever patterns resonate with you the most. The important thing is that you are progressively learning to become adept at finding celestial wonders easily and as quickly as possible.

If you have not already tried some of these star pattern jumps, you are encouraged to do so as soon as possible. Even an hour or so with binoculars, following the steps mentioned for each object, would be a valuable aid to your memory. It will also help turn this intellectual exercise into a practical one that will help you each time you go out to observe. If it is autumn or winter, then you can try the star pattern jumps given for M31, M33, and M81/M82. If it is springtime try M51, M101, and in summer try locating M83 and M104.

Also, in Fig. 5.11a there is another easy way to find the galaxy that has not been discussed. It is the magnitude 8.4 galaxy, M106. Why not use Fig. 5.11a along with whatever star charts that you own and plot out a path to this destination. On your next observing session, give it a try and be sure to record your own impressions of this galaxy. Interacting with what you read in such a practical way will help maintain your excitement for visual astronomy and build confidence in your ability to find these easier objects. You will need this confidence as we move on to much more challenging celestial treasures.

For now, let's continue with a few more easy galaxy targets that will provide a good bridge to the next section, where we will discuss much more challenging ones. Refer again to Fig. 5.11a, and find the 2.5 magnitude star Beta (β) Ursae Majoris (Merak). Just 1.5° southeast of Merak is the galaxy **M108** (NGC 3556), discovered by Pierre Méchain on February 16, 1781, along with the Owl Nebula, **M97** (NGC 3587). Both objects are tenth magnitude and are particularly interesting to observe, since they will both fit into the view of a wide field eyepiece. With this pair you can observe a galaxy and a planetary nebula at the same time. The nebula is much closer, of course, an estimated distance of 2,000–4,000 light years away, while the galaxy is 23 million light years distant. (Some estimates are as high as 46 million.)

Small telescopes will easily display this duo, and in a 6-in. 'scope more details become apparent. The reason why M97 is called the Owl Nebula becomes clear as the two dark spots on this circular shaped planetary resemble the piercing eyes of an owl. There is a central 16th magnitude star in the Owl Nebula, but to see it will require a very large 'scope (perhaps upwards of 16 in. [406 mm]). The M108 galaxy appears long and light colored, looming like a large barge through space. It is not the most striking galaxy in of itself, but the thrill of viewing it comes in part because of its pairing with M97 (Fig. 5.14).

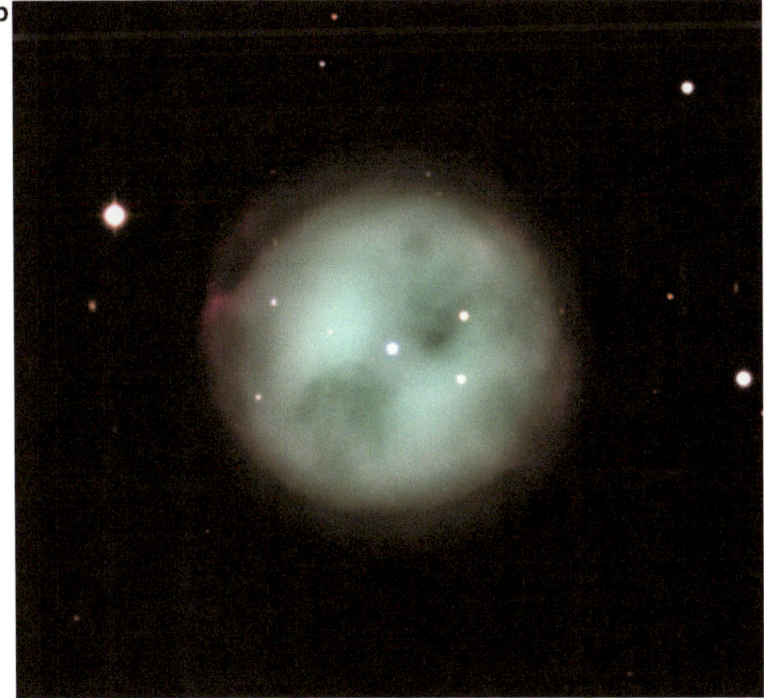

Fig. 5.14 (a) The galaxy M108 in Ursa Major. (Image courtesy of NOAO/AURA/NSF.)
(b) The Owl Nebula (M97), near M108 in Ursa Major. (Image courtesy of NOAO/AURA/NSF)

Next up is another galaxy also discovered by Méchain in early 1781—**M109** (NGC 3992). When you compare it to M108, you'll probably note that it looks clearer and sharper. It is magnitude 9.8 but has a lower surface brightness than M108 and is slightly smaller in angular size. So then what accounts for its clarity? Instead of being almost edge-on, like M108, M109 is much more tilted towards our line of sight, appearing almost face-on (Fig. 5.15a).

This beautiful barred spiral galaxy is a staggering 60 million light years distant, according to recent measurements. M109 also has three faint companion galaxies within 14′ of it. Astronomers have found evidence of dark matter surrounding this group of galaxies. Dark matter has in recent years gained more and more attention, and much more time and research will be needed before astronomers fully understand it. The three companion galaxies to M109 are UGC 6969, UGC 6940, UGC 6923, and each are magnitude 14.6, 16.7, and 13.4, respectively. Although much too faint for all but the largest amateur telescopes, the latter one may be possible (UGC 6923); however it is only 4′ away from a bright 8.6 magnitude star. The brightness of that star will likely prevent you from seeing it. Use a 12.5-in. (317 mm) or larger 'scope, and try moving the star just out of your field of view to see if UGC 6923 can be detected.

To find M109, look just 38′ (about ½°) east-southeast of the 2.5 magnitude star Gamma (γ) Ursae Majoris. Fortunately, small telescopes will easily reveal not only M109, but also another spiral galaxy nearby M109 that is designated **NGC 3953** (magnitude 10.1). Like M109, NGC 3953 is about 60 million light years away. Along with M109 it is a part of a larger group of galaxies in the Ursa Major cluster. It lies just 1° south-southwest of M109. NGC 3953 has had two known supernovae, one in 2001 and one in 2006. Both M109 and NGC 3953 are shown in Fig. 5.11a. An enlarged view of M109 with its three faint UGC (*Uppsala General Catalogue of Galaxies*) companions and NGC 3953 are shown in Fig. 5.15b.

The next group of galaxies we will consider are a part of the Leo II Group in the constellation of Leo. This galaxy group, along with our own Local Group and others, are a part of the large Local Supercluster (also called the Virgo Supercluster). On October 12, 1773, Charles Messier discovered his ninth comet, and that same comet passed right by two galaxies that he would later discover. At the time, the bright light of the comet prevented him from seeing the two "nebulae" that he later described as faint. So it was that on March 1, 1780, he finally saw both M65 and M66 in Leo. **M65** (NGC 3623), **M66** (NGC 3627), and **NGC 3628** are a part of what is called the Leo Triplet, 30 million light years distant. Astronomers have concluded from spectroscopic evidence that these galaxies had tidal interactions with each other many, many years ago. If you enjoyed viewing the combination of M108 with the nearby Owl Nebula, you are sure to love the sight of these three galaxies together in the same field of view. All three galaxies are approximately tenth magnitude. The two Messier-numbered galaxies are near face-on. The other galaxy presents an edge-on view (Fig. 5.16).

It's possible to see them in large binoculars, perhaps 10×50 mm or larger. As in the case of our previous examples, although visible in smaller apertures, 'scopes upwards of 6 in. will reveal the most detail in this trio. The dust lane in NGC 3628 becomes especially interesting in larger apertures.

Fig. 5.15 (**a**) M109 in Ursa Major. (Image courtesy of NOAO/AURA/NSF.) (**b**) Enlarged area showing the M109 and its companion galaxies. (Star chart courtesy of *Sky Tools 3* at www.skyhound.com)

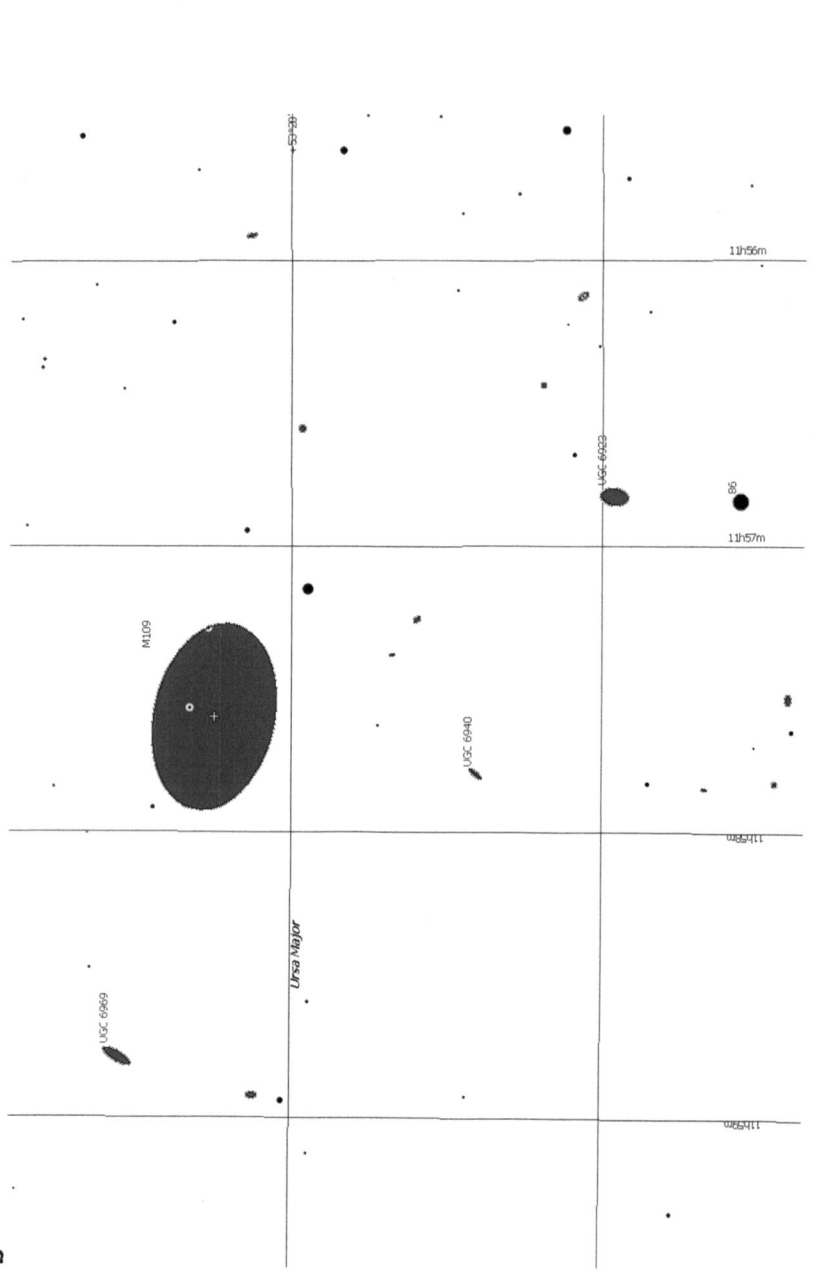

Fig. 5.15 (continued)

Fig. 5.16 The Leo Triplet, including galaxies M66 (*lower left*), M65 (*lower right*), and the edge-on galaxy NGC 3628. (Image courtesy of REU program/NOAO/AURA/NSF)

To locate this group, start with the east to find 1.3 magnitude star Regulus (Alpha [α] Leonis), and move 16° east to Theta (θ) Leonis. Just 2° south of θ Leonis is a group of six stars that form a fishhook pattern. You'll find the catch of galaxies within 1° of the westernmost star (fifth magnitude 73 Leonis) of the hook. Take a look at Fig. 5.17 to help you find your way there.

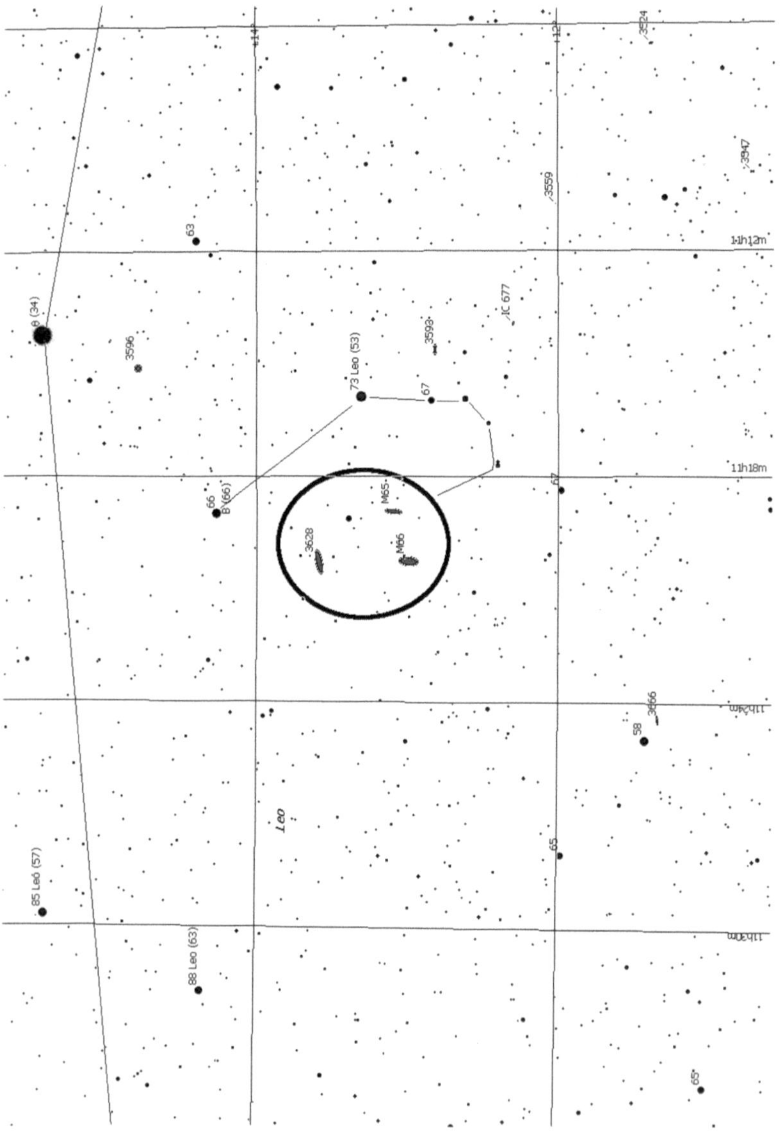

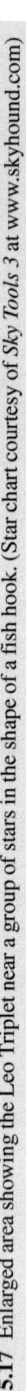

Fig. 5.17 Enlarged area showing the Leo Triplet near a group of stars in the shape of a fish hook. (Star chart courtesy of *Sky Tools 3* at www.skyhound.com)

Our next stop is the Leo I Group of galaxies (about 34 million light years distant) just south of the star 52 Leonis. Here within the same field of view as this striking blue star are two galaxies, elliptical galaxy **NGC 3377** and face-on spiral galaxy **NGC 3367**. NGC 3377 extends about 40,000 light years in size, and like many galaxies, it contains a supermassive black hole with the equivalent mass of 100 million suns. It has a close companion galaxy (NGC 3377A) that is magnitude 13.6 and has a magnitude 14.7 star superimposed on its eastern side (Fig. 5.18).

The spiral galaxy NGC 3367 extends for 80,000 light years and is actually the fainter of the two at magnitude 11.5, while the elliptical NGC 3377 appears much brighter at magnitude 10.4. Out of all the galaxies covered so far, these will appear to be the faintest. This is a good bridge into our next section because the galaxies featured have become progressively fainter so as to gradually build your tolerance and appreciation for the fainter objects. Even if your interest in galaxies is not as great as other deep sky objects, by going through each of these your eyes will gradually learn to pick up on more details in all deep sky objects. Also, all those objects you may have once viewed as faint will become seemingly brighter.

In a 6-in. (152 mm) telescope you will be able to see both of these galaxies. The elliptical galaxy NGC 3377 will appear brighter and the spiral NGC 3367 will be quite faint. In a 10-in. (254 mm) or larger this difference will be more pronounced; in fact the elliptical seems to be brighter, more elongated, and more bulged to the point that it seems like it is the spiral. The actual spiral is face-on, and because of its fainter magnitude and much smaller angular size, its spiral arms are not clear. Nevertheless the two of them together present an interesting and enjoyable observing challenge. Be sure to use averted vision and see if you can pick up one or more details in the spiral NGC 3367.

There are many other galaxies that you should explore in the Leo I Group. Thirty years ago it was discovered that NGC 3384 and M96 have an immense intergalactic H I (neutral hydrogen) cloud positioned halfway between them. This cloud of neutral hydrogen is about 652,000 light years in diameter. Some astronomers believe this is left over gas from which the many galaxies in this group were formed. The galaxy trio of the lenticular galaxy **NGC 3384**, the spiral **NGC 3389**, and the elliptical galaxy **M105** (NGC 3379) form another spectacular group that all can be seen at one time in a wide field of view. The galaxies are magnitude 9.9, 11.9, and 9.3, respectively. They are positioned relative to each other differently than the Leo Triplet, and are just as impressive to observe. Just 1° south-southwest of this trio is 9.3 magnitude **M96**—a barred spiral, and then 35′ west of it is **M95**, another magnitude 9.7 barred spiral. Figure 5.19 show an enlarged map of this group of galaxies that are tied together.

What if you just cannot seem to be able to find one of the galaxies discussed? Be sure that your eyes are fully dark adapted, and always try averted vision in addition to direct vision. You should also experiment with averting your eyes in different directions, such as looking up instead of to the side. Also, try not to rush as you jump from one pattern of stars to another. Take your time and get familiar with the

Fig. 5.18 Two nearby galaxies in the Leo I Group. NGC 3377 is the elliptical galaxy in the center, and NGC 3367 is the spiral shown at *lower left*. (Image by the author)

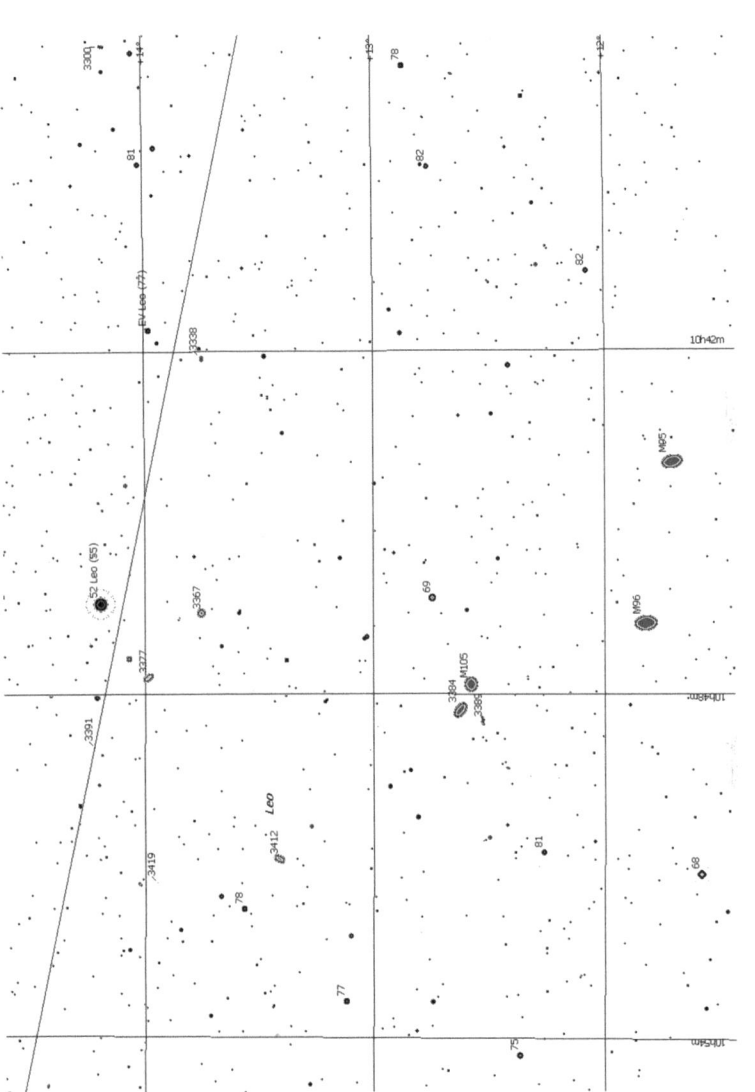

Fig. 5.19 Enlarged area showing the Leo I Group of galaxies. (Star chart courtesy of *Sky Tools 3* at www.skyhound.com)

layout of the stars in the area. Leaving the star field and coming back to it once or twice ensures that you have the correct stars in view. Sometimes it is just a matter of poor seeing conditions that inhibit views of a galaxy. For especially faint objects, give your eyes several minutes of viewing to let them accumulate details. Eventually your eyes will sort of tap you on the shoulder and say "There it is!".

What are some things to look out for as you examine these galactic cities containing billions of stars? The most obvious would be the size of the galaxy and how it compares to others you've seen. Then try to discern its shape. Is it a spiral, barred spiral, elliptical, lenticular, irregular, or peculiar galaxy? What is the angle of view for this galaxy—edge on, face-on, or perhaps tilted to our line of sight? Look for dust lanes, spiral arms, knots, or other oddities. Are there any striking stars nearby in the field of view? What color are those stars? Sometimes there are stars that are superimposed over the galaxy, meaning the stars are really in our own galaxy but are in your line of sight towards the external galaxy that you are viewing.

One of the most fascinating observations that can be made in a galaxy is that of a supernova. Professional astronomers use automated search programs run by computerized telescopes to actively monitor galaxies that are likely candidates for this kind of activity. Amateur astronomers have also discovered many supernovae by the comparison of photographs or CCD images of a galaxy. Images were compared to earlier ones and then the appearance of a "star" that was not there before was noted.

However, many amateur discoveries have been made by visual inspection of a galaxy. For example, consider a veteran amateur astronomer living in Australia, Robert Evans. His visual prowess for immediately recognizing anything that seems out of place in a galaxy has become well known over the last several decades. His first discovery was in 1981 with a 10-in. reflector. After ten more discoveries with his 10-in., he later began using a 16-in. that he acquired through funding from a government-sponsored research program. Since then he has discovered more than 40 supernovae, and at least 39 of them have been visual observations through a telescope with the aid of star charts. In the November 1989 issue of *Astronomy* magazine (p. 97) Robert Evans describes a night of observing (in 1986) when he visually inspected 338 galaxies in one night. It turned out that a supernova was visible in the very last galaxy that he observed that night. Was it just luck? No. It was a combination of patience, dedication, and tenacity.

There are numerous other easy to find (and some not so easy to find) galaxies besides the ones already discussed. A good place to continue is with the rest of the galaxies in the Messier catalog. You can find them all listed in a star atlas such as *Uranometria 2000.0*. The next step would be to locate the hundreds of galaxies plotted in *Sky Atlas 2000.0* that are within range of a medium-aperture 'scope. Remember, too, the star atlas *Uranometria 2000.0* contains over 25,000 total galaxies brighter than 15th magnitude, so there are plenty to choose from if you are able to use a large 'scope. Before moving on to the next section, which presents some challenging galaxies, take a few minutes to reflect on what you've learned by using the questions below.

Helpful Things to Consider

What bright stars are in the vicinity of the galaxy I wish to target? After reviewing a star chart or atlas, are there any star patterns that can serve as guideposts to my destination?

What other deep sky objects (including other galaxies) are nearby the initial one I observed? How do they compare with my original target?

What are my initial impressions of a galaxy compared with my impressions after allowing sufficient time to pass so that my eyes can accumulate more details?

What is my preferred magnification on this object?

A Few Challenging Galaxies

Now it's time to branch out to a few targets that you may find a bit more challenging to find at first. With lots of practice, you'll be pointing your 'scope at objects almost without thinking about it. Our first destination will be in the Virgo Cluster of galaxies. Interacting galaxy pair **NGC 5427** and **NGC 5426** were first discovered by William Herschel in 1785. Years later it was included in the 1959 *Catalogue of Interacting Galaxies* by Vorontsov-Velyaminov, and again in 1996 it was included by Halton Arp as Arp 271 in his *Atlas of Peculiar Galaxies*.

This peculiar pair is located in the Virgo constellation and is 87 million light years distant. Together the interacting pair covers a distance of over 130,000 light years. NGC 5427 is a face-on, magnitude 11.4 Seyfert galaxy and NGC 5426 is an elongated spiral galaxy of magnitude 12.1. These galaxies can be clearly seen interacting even in visible light, connected by a bridge of stars and ionized gas. The spiral on the bottom of the image (see Fig. 5.20) is NGC 5426 and is in front of the other (slightly closer to us), and as it is slightly more massive, it is causing more of an effect on NGC 5427 (in velocity and rotation curve). Similar to what is happening with our galaxy and the Andromeda Galaxy, these two galaxies are engaged in a dance that seems to be much older than ours. Whether or not they will end up in a complete collision is still undetermined.

To locate this dynamic pair, begin with Spica in Virgo. A 13.5° distance stretches between Spica and fourth magnitude Iota (ι) Virginis (Syrma). The galaxy pair sits between two patterns of stars. One is a rectangular shape just northeast of Spica with the fifth magnitude star 82 Virginis on its western corner. The other pattern is a triangle formed by Syrma, Kappa (κ) Virginis, and 95 Virginis. The NGC 5427 pair is 3° due west of Syrma.

You will need at least a 6-in. 'scope to make out this pair and the distinction between the two. A 10-in. will begin to show indications of the spiral nature of NGC 5427, while NGC 5426 remains much smaller in appearance. Figure 5.21a, b

Fig. 5.20 Two interacting galaxies 87 million light years away in the Virgo Cluster. NGC 5427 is the face-on spiral shown above its companion, NGC 5426. (Image by the author)

should be very helpful in locating this target. Begin first with binoculars and make sure that you recognize the patterns of stars from the chart. Once they become familiar, move on to your finderscope. In the finderscope center the star Syrma and then move slowly 1° west, and center the tiny triangle of stars in the finderscope. From here you should switch to the eyepiece. If you are using an eyepiece with about a 1° field of view, you will only need to gently move your 'scope two eyepiece fields over to the west in order to find this galaxy pair.

Now we turn our attention to the circumpolar constellation of Draco, to take a look at a spiral galaxy different from the others discussed so far. **NGC 5907** (sometimes called the Splinter Galaxy) is a very long galaxy, razor thin in appearance. It was discovered by William Herschel on May 5, 1788. The Splinter Galaxy lies at least 35 million light years away (although some estimate 45 million), and its size stretches across 124,000 light years. Surrounding it is a faint stellar halo that is thought to perhaps have been acquired from either a close brush with an elliptical galaxy or a dwarf companion galaxy long ago. At magnitude 10.3 it is easily seen in small telescopes. In larger apertures the galaxy's long, thin appearance is more pronounced (Fig. 5.22).

To see the Splinter Galaxy begin with second magnitude Alkaid (Eta [η] Ursae Majoris), which points directly northeast to the star Edasich located in Draco (Iota [ι] Draconis) about 17° away. From Edasich, the Splinter lies just 3° south-

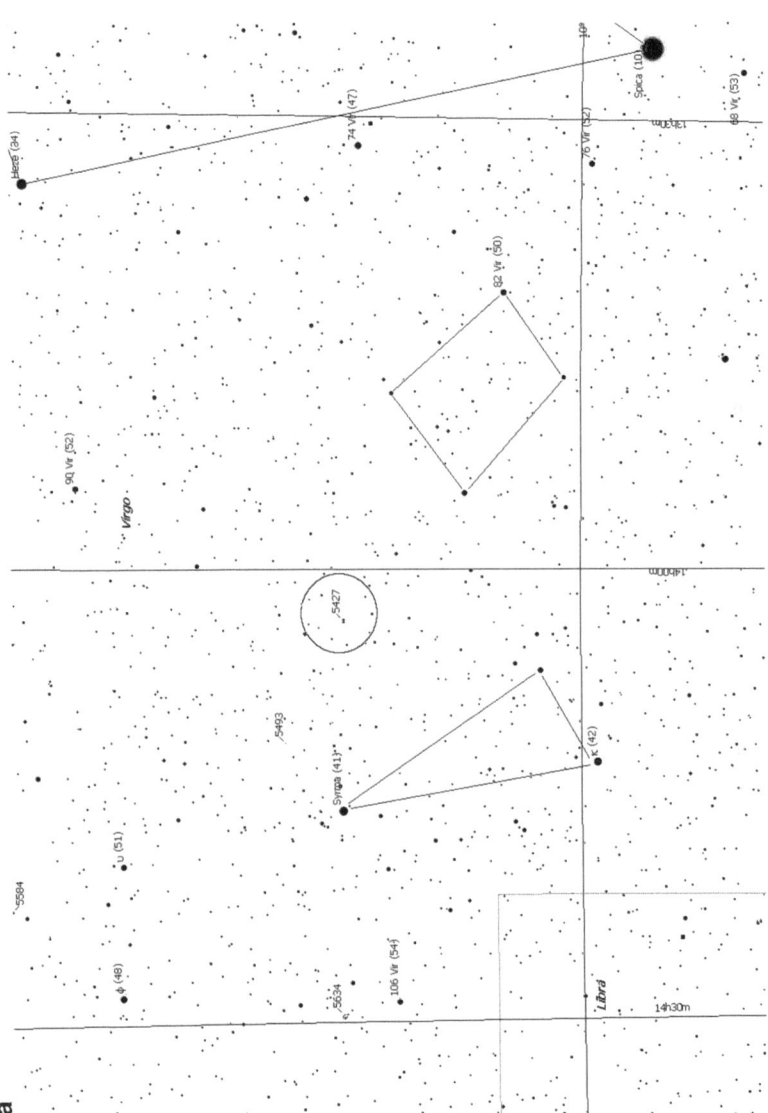

Fig. 5.21 (**a**) Broad area near NGC 5427 and NGC 5426. The bright star Spica dominates the general area, and several other stars serve as helpful guideposts to locate these galaxies. (Star chart courtesy of *Sky Tools 3* at www.skyhound.com.) (**b**) Enlarged area near NGC 5427 and NGC 5426. Note the nearby triangle of stars pointing the way, located between the star named Syrma and the two galaxies. (Star chart courtesy of *Sky Tools 3* at www.skyhound.com)

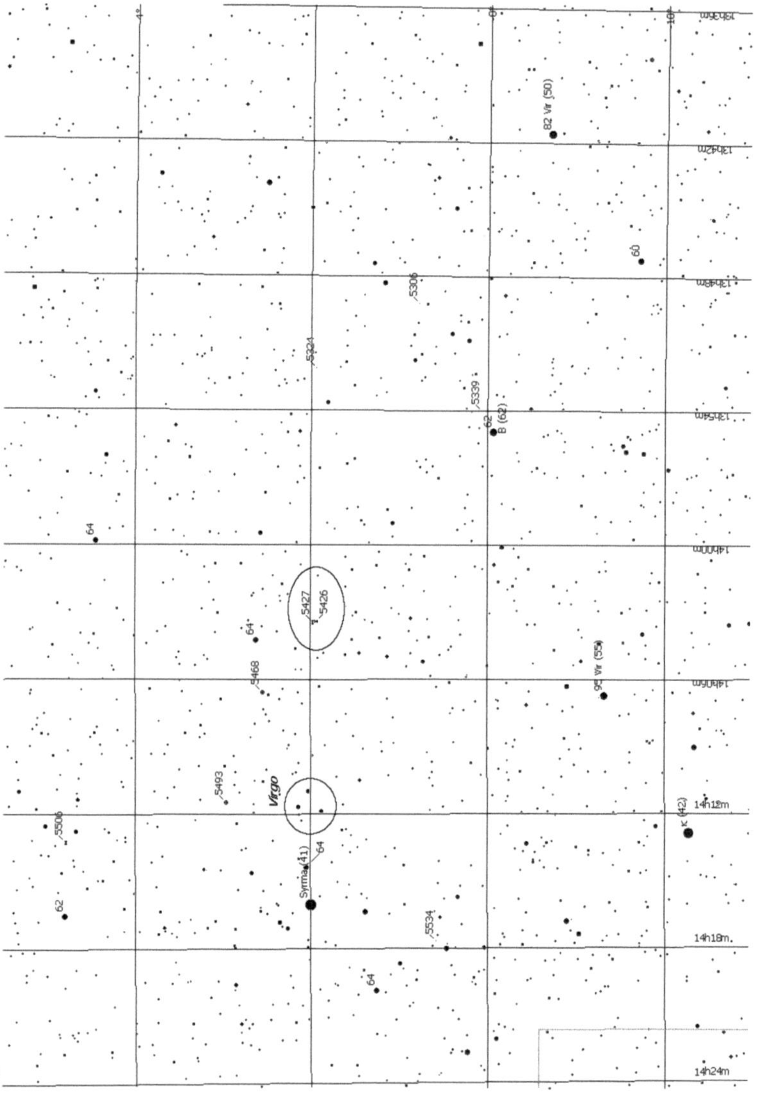

Fig. 5.21 (continued)

Fig. 5.22 NGC 5907 in Draco. (Image by the author)

southwest. Use Fig. 5.23 to find it. What star patterns will you choose to help you arrive at NGC 5907? While you are in the area, take some time to observe some of the other galaxies nearby, such as M102 (NGC 5866). Some believe that M102 is a mistaken duplication of M101, but many astronomers believe that NGC 5866 is still the most viable explanation for what Messier later cataloged as M102. M102 is just 1.5° to the southwest of the Splinter Galaxy.

Also in Draco is a very pretty galaxy with a bright core. **NGC 5987** is just over 2° southeast of Edasich, and it shines dimly at magnitude 11.7. It is a very large spiral that extends 260,000 light years across and is located very far away. Very near this galaxy, there is a bright 6.5 magnitude star just 11′ to the southwest. That star (designated as HD 140117 in the Henry Draper Catalogue) speeds towards us at about 8 km per second. Together with NGC 5987 this makes a great contrasting visual pair, since the star is only 600 light years away while the galaxy is 142 million light years away. Figure 5.24 shows the area centered on NGC 5987.

The faint magnitude of this galaxy makes it tough to see a lot of detail. However the fact that it is within the same field of view as the 6.5 magnitude star will give you an edge in locating it. Whether you are using a 6-in. or even a telescope of over 10 in. (254 mm) you will for the most part only see the bright core of this galaxy. In images of this galaxy two dust lanes can be clearly seen. A very large amateur telescope is likely to show some of this kind of detail (Fig. 5.25).

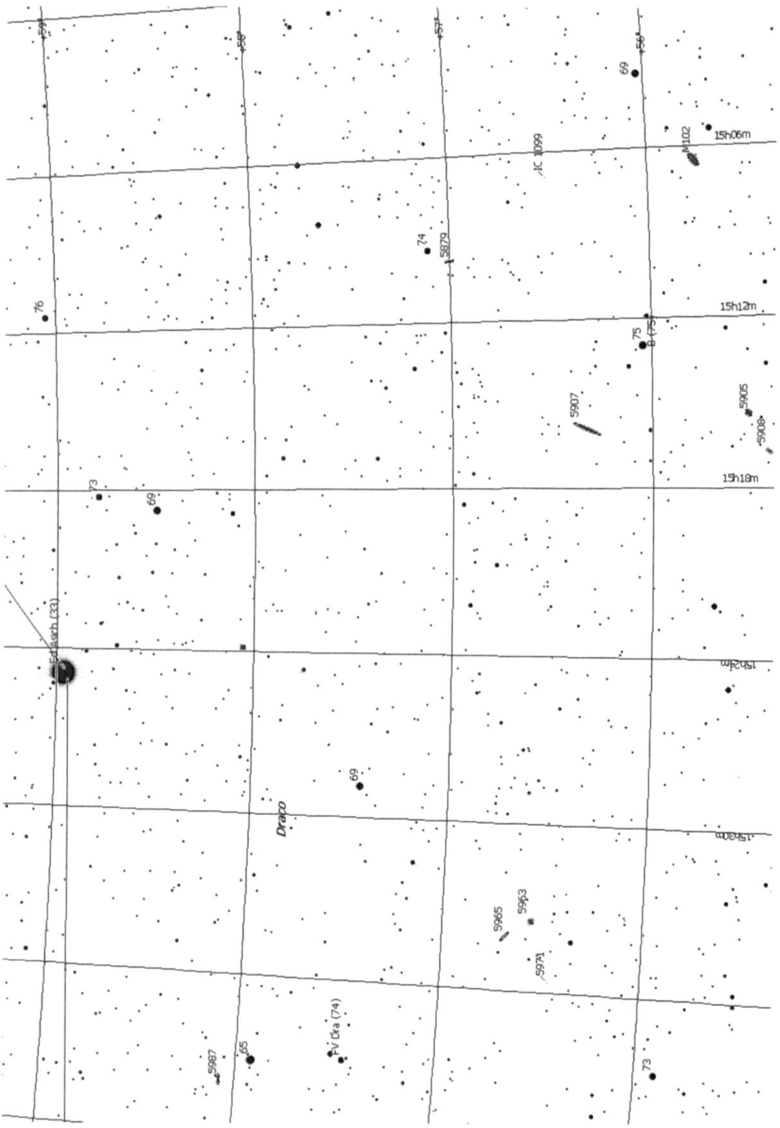

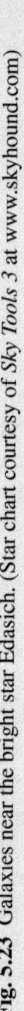

Fig. 5.23 Galaxies near the bright star Edasich. (Star chart courtesy of *Sky Tools 3* at www.skyhound.com)

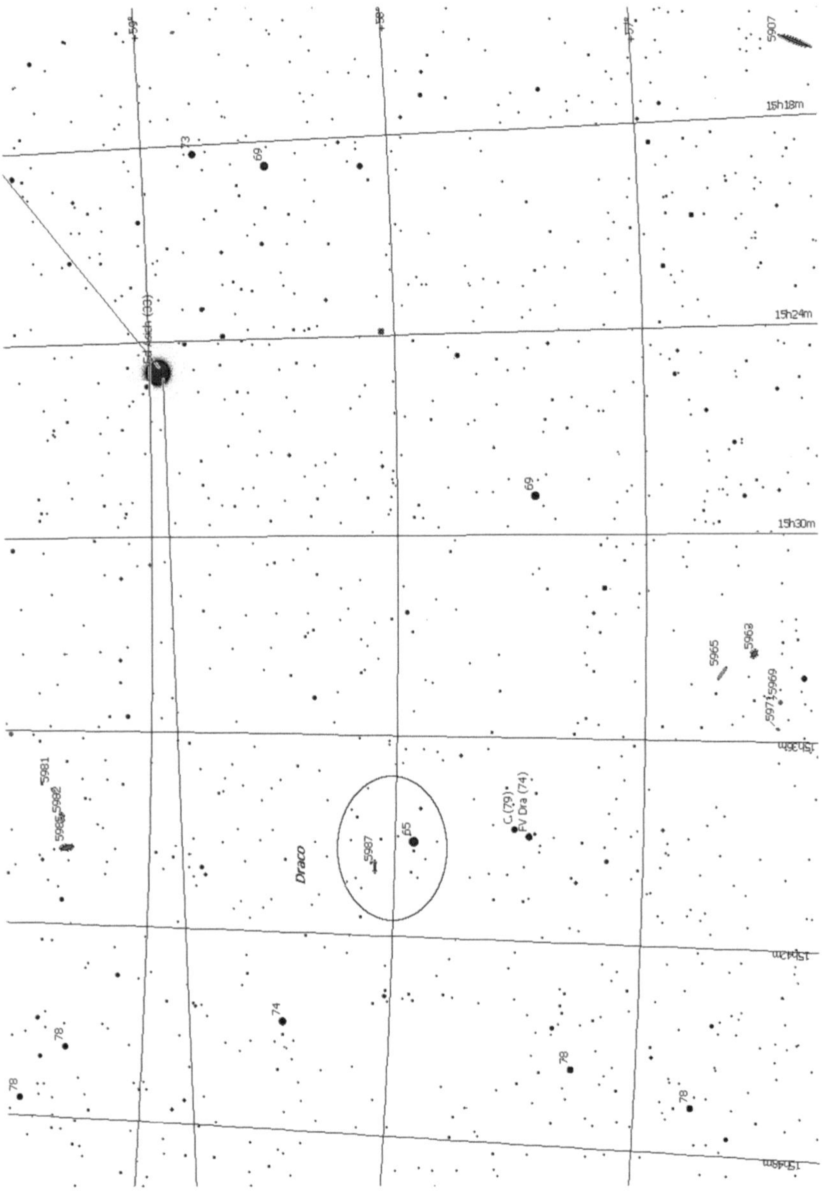

Fig. 5.24 Enlarged area near the spiral galaxy NGC 5987. (Star chart courtesy of *Sky Tools 3* at www.skyhound.com)

Fig. 5.25 NGC 5987 in Draco. (Image by the author)

For an even greater challenge try **NGC 5965** and **NGC 5963**, a dimmer galaxy pair 1.5° southwest of NGC 5987. NGC 5965 is magnitude 11.7, 160 million light years distant and NGC 5963 is magnitude 12.5, about 35 million light years distant. NGC 5963 is small, at only 35,000 light years across, whereas its neighbor is more than seven times that size. A 10-in. 'scope will show them, but even if you have a smaller aperture see if you can make them out. They should be within range of a medium-aperture telescope although tough because of their lower surface brightness. In a 12.5-in. (317 mm) 'scope the two appear close enough to be intertwining, but this is just an illusion because the two galaxies are not physically related.

These last two challenging targets may have pushed your observing skills to the limit. If you had a little trouble locating them do not despair—that's natural and to be expected. Focus on the galaxies that you did locate successfully and learn from the process you followed to do so. Later you can return to the more challenging galaxies after your mind has had a chance to absorb and reflect on the process. Our next object is a much brighter galaxy that can be found in the constellation of Pegasus during autumn and winter.

NGC 7331 is a magnitude 9.5 spiral galaxy about 46 million light years away in Pegasus and was discovered by William Herschel on September 5, 1874. It is massive in size, extending 140,000 light years and is inclined towards us at about 77°. In the infrared wavelength the Spitzer Space Telescope has detected a large ring of gas made up of carbon monoxide, neutral hydrogen, as well as stellar remains from

Fig. 5.26 NGC 7331 in Pegasus. (Image by the author)

stars. Astronomers suspect it may have interacted with a nearby dwarf galaxy. NGC 7331 is the dominant member of the NGC 7331 galaxy group and is loosely tied to at least nine other galaxies within 1 million light years of it (Fig. 5.26).

This large spiral can be found by first locating the Great Square of Pegasus and centering on the star on its northwestern corner, 2.5 magnitude Beta (β) Pegasi. That star forms a triangle with Lambda (λ) Pegasi and Eta (η) Pegasi. From the third magnitude star η Pegasi, go north 4.5° to find NGC 7331. You should just about be able to detect it in binoculars, and it is clearly seen in small telescopes. In a medium-size 'scope this galaxy shows a bright center and a dust lane. Its appearance is reminiscent of the Andromeda Galaxy, although its angular size is smaller. Figure 5.27a, b present a broad view and an enlarged view of the area near NGC 7331.

Another interesting galaxy in Pegasus is **NGC 7814**. It was discovered on October 8, 1784, by William Herschel. It resides at a distance of about 48 million light years and is a large spiral galaxy that extends 86,000 light years. The system of globular star clusters in NGC 7814 is similar to the one in our Milky Way Galaxy, and consists of approximately 200 clusters. In photographs this galaxy's dust lane is very distinct, but this feature can be seen visually only through very large amateur telescopes. However, a medium 'scope will show its bright and bulged core (Fig. 5.28).

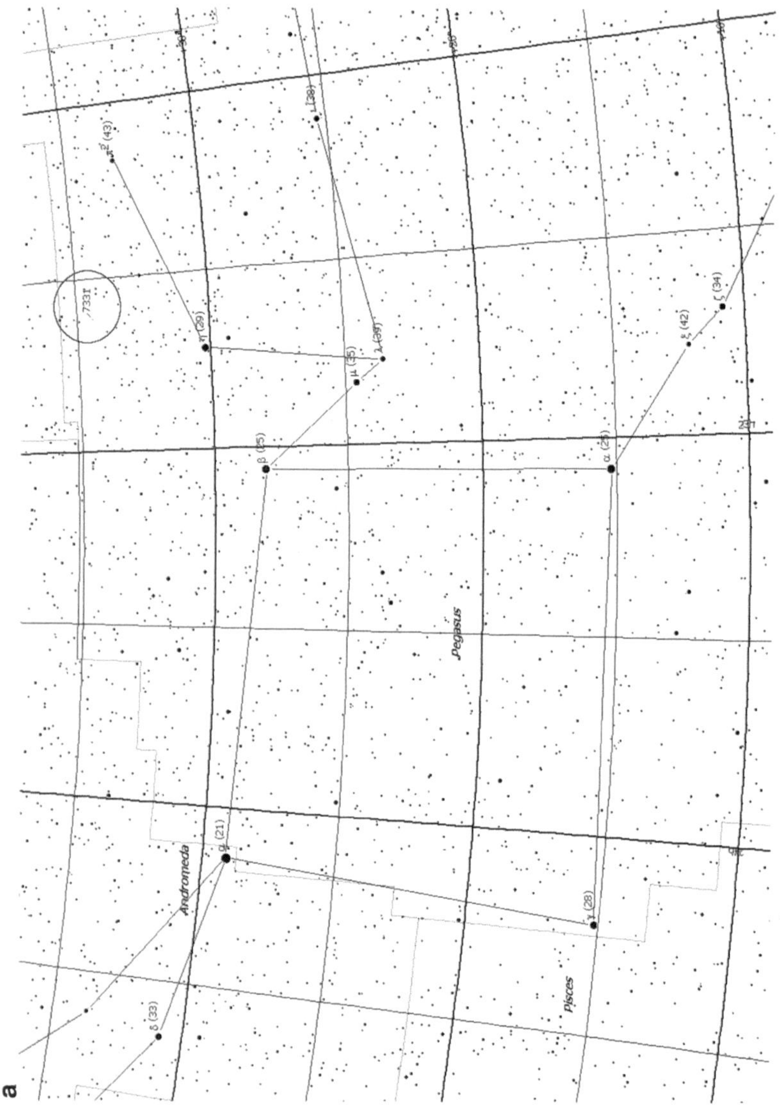

Fig. 5.27 (**a**) The Great Square of Pegasus. (Star chart courtesy of *Sky Tools 3* at www.skyhound.com.) (**b**) Enlarged area near NGC 7331 in Pegasus. (Star chart courtesy of *Sky Tools 3* at www.skyhound.com)

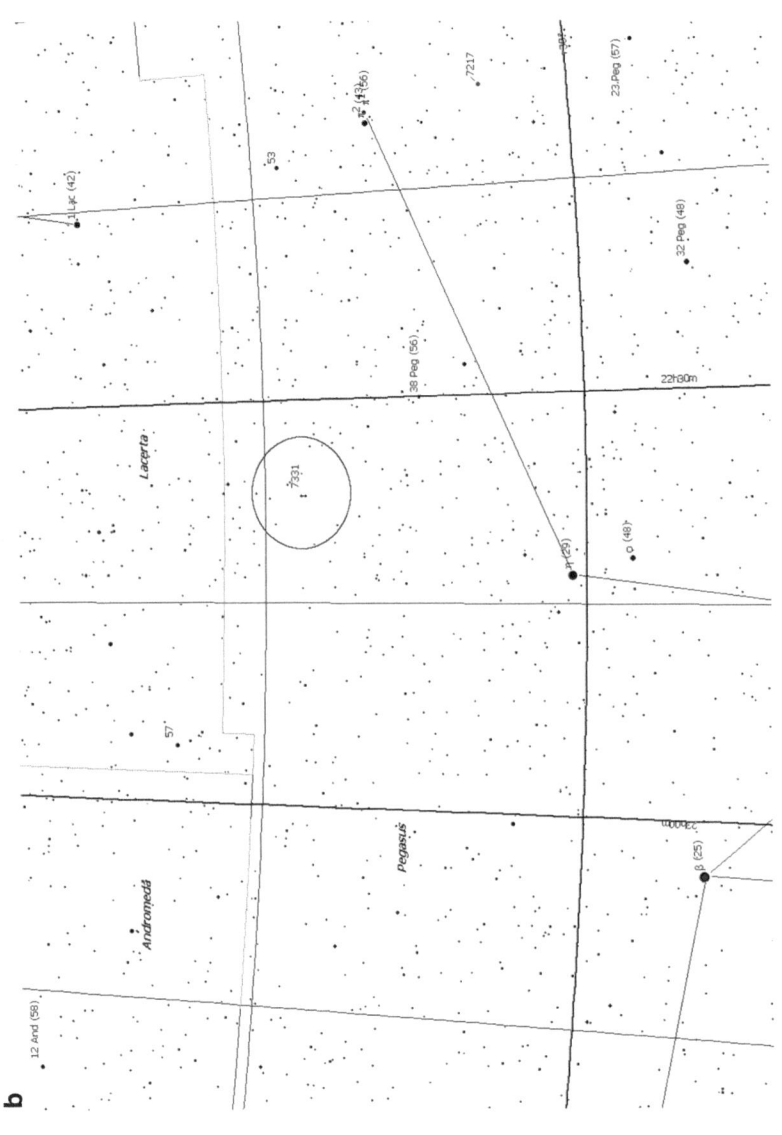

Fig. 5.27 (continued)

Fig. 5.28 NGC 7814 in Pegasus. (Image by the author)

The best way to find NGC 7814 is to begin with the third magnitude Gamma (γ) Pegasi, the star that sits on the southeastern corner of the Great Square of Pegasus. Just 2.5° west-southwest is a triangle of seventh and eighth magnitude stars. NGC 7814 is only 1.5° north of that triangle. A seventh magnitude star sits within 10′ of the galaxy's northwest corner. See Fig. 5.29a, b for additional help in locating this galaxy.

Our final galaxy to discuss in this chapter is **NGC 681**, an estimated 75–80 light years away in the constellation of Cetus; this object was discovered by William Herschel on November 28, 1785. Although at magnitude 12.0, it is within reach of

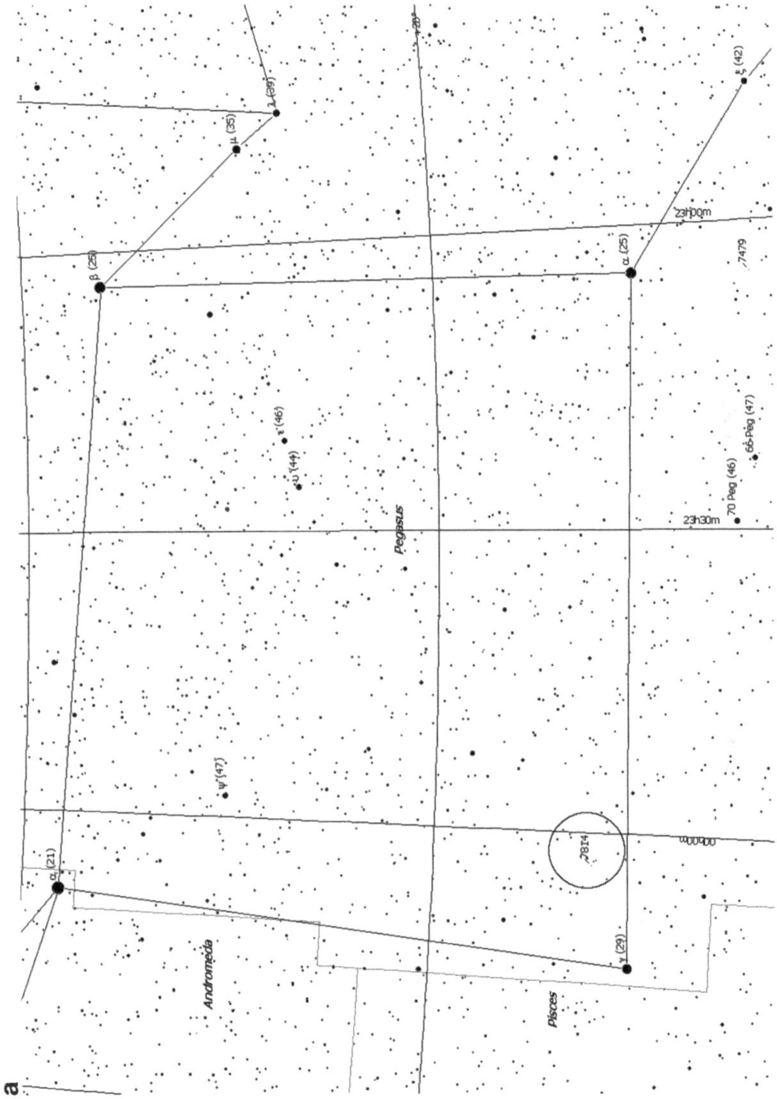

Fig. 5.29 (**a**) Area near NGC 7814. (Star chart courtesy of *Sky Tools 3* at www.skyhound.com) (**b**) Enlarged area near NGC 7814. (Star chart courtesy of *Sky Tools 3* at www.skyhound.com)

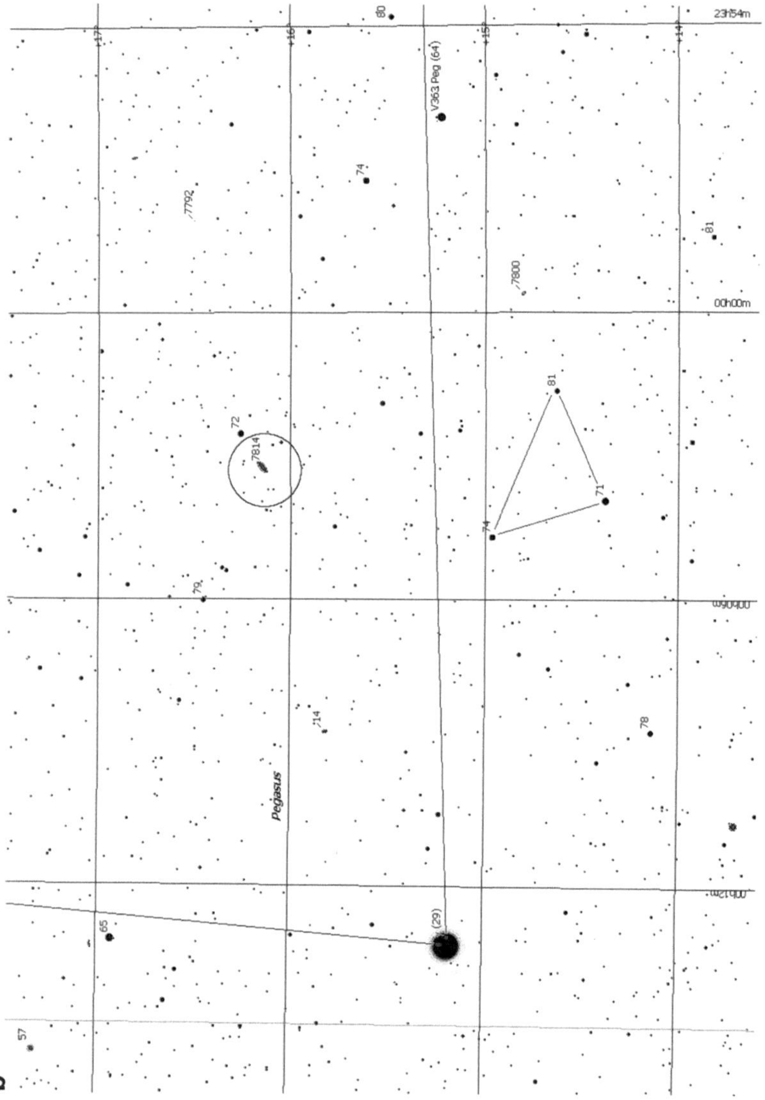

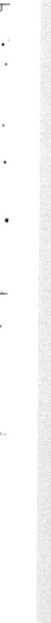

Fig. 5.29 (continued)

Fig. 5.30 NGC 681 in Cetus. (Image by the author)

medium-size telescopes and will appear brighter than its magnitude seems to indicate. It has been compared to a miniature Sombrero Galaxy because of its striking visual similarity to M104. At medium power the dust lane can be clearly seen under dark, transparent skies (Fig. 5.30).

You'll find NGC 681 only ½° west of Zeta (ζ) Ceti. You may recall that in the last chapter we were just one constellation over to the west, searching for the Helix Nebula in Aquarius. With a newfound appreciation for much fainter objects (galaxies), you may want to return to the Helix for another look and see how your visual estimation of this object has changed. See Fig. 5.31a, b for both a broad and magnified view of the area near NGC 681. There are many other celestial sights in this constellation, so be sure to browse.

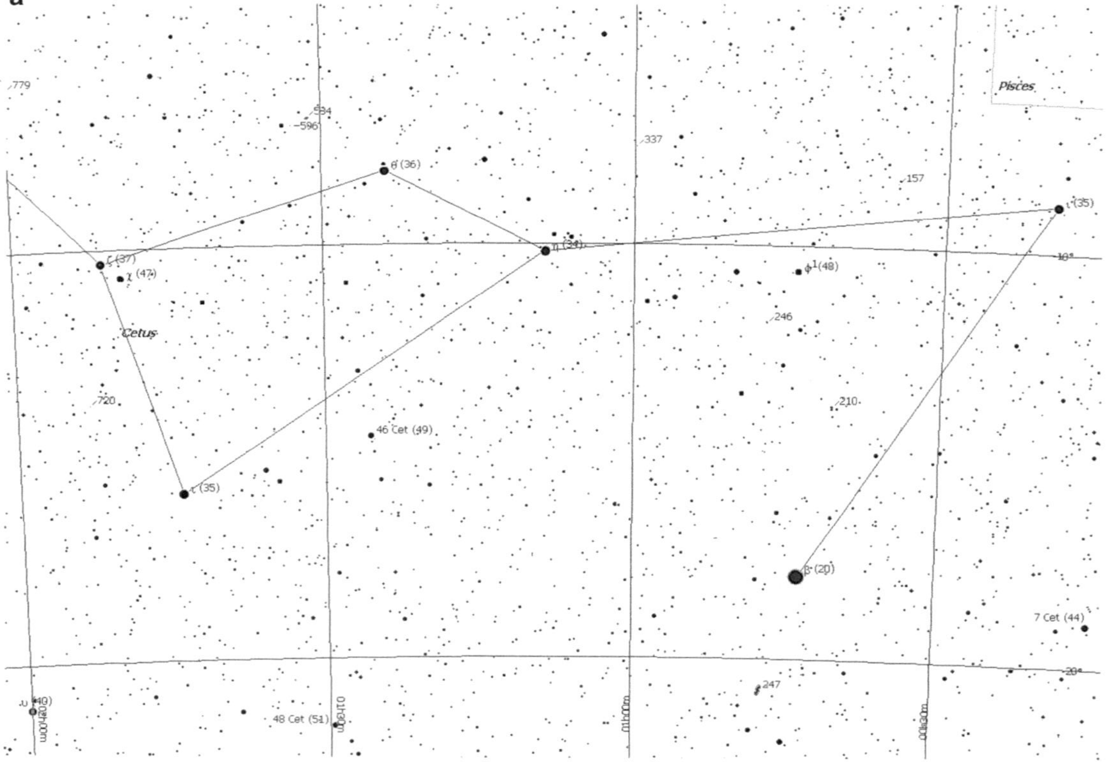

Fig. 5.31 (**a**) The Cetus constellation. NGC 681 is near the star labeled Zeta (ζ) Ceti. (Star chart courtesy of *Sky Tools 3* at www.skyhound.com) (**b**) Enlarged area showing NGC 681 is near Zeta (ζ) Ceti and Chi (χ) Ceti. (Star chart courtesy of *Sky Tools 3* at www.skyhound.com)

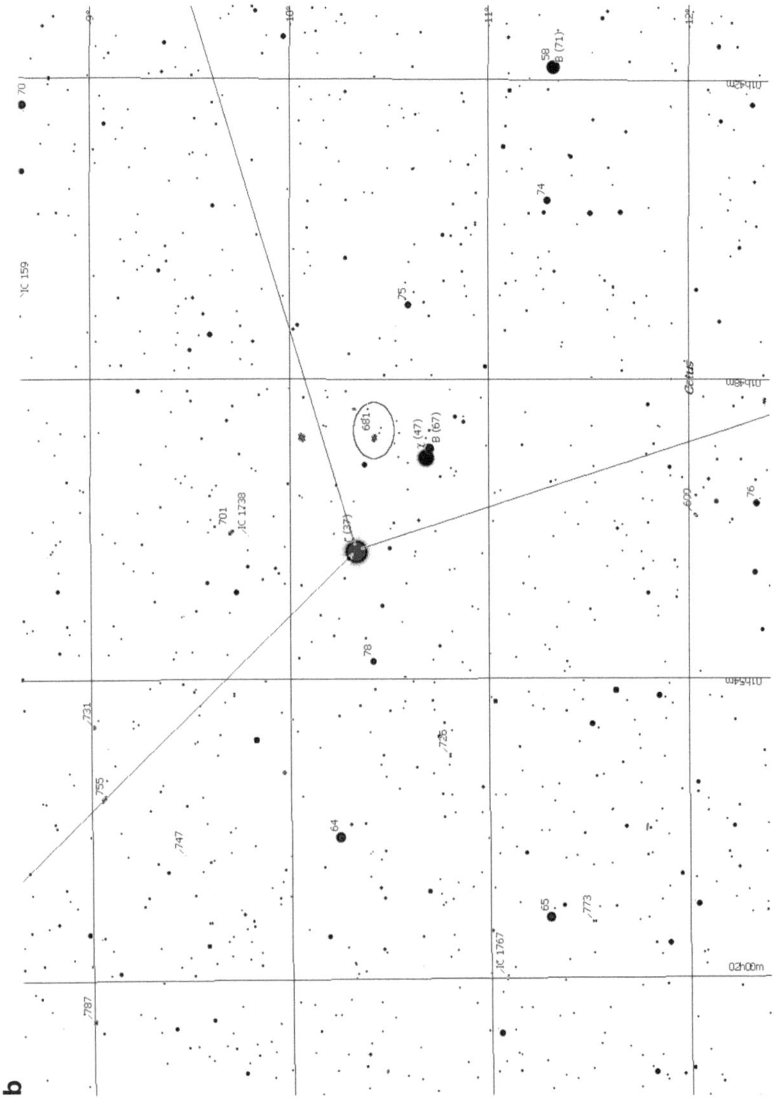

Fig. 5.31 (continued)

Keep Expectations Reasonable

Visually observing galaxies millions of light years away from Earth is a thrilling experience. Expect to be awestruck as you find distant objects you may at one time felt were too difficult to find. Build your confidence and ability to locate galaxies, or any object for that matter, by taking your search step by step, and retracing those steps if necessary. At the same time do not expect perfection from yourself, and do not expect to find everything on the first attempt. Some deep sky objects will just have to wait for the next night of clear skies. Learn to take your time and appreciate the galaxies that you *have* been able to locate during a night of observing.

Generally speaking you should begin each search for your target object by surveying the area with binoculars. If this is something that you already did, as suggested in earlier chapters, you probably took note of several fuzzy objects that turned out to be star clusters, nebulae, and even galaxies. Now, on the return trip, your focus should be not only to recognize these fuzzy celestial treasures but to see them in relation to star patterns. A simple star pattern seen in binoculars leads to a pattern that can also be recognized in your finderscope. Then, always begin with the widest field eyepiece you own, and afterwards adjust the magnification to the best view for each particular object. Do not expect superb views of galaxies under high magnification. Higher magnifications can reveal subtle details, but remember that the galaxy will appear dimmer. Therefore, low and medium powers are usually a better choice for prolonged viewing of a galaxy.

Although the objects that are presented here are in an order that will help you grow from one target to the next, you can feel free to observe in any order you like. As stated previously, you are always the captain of your stellar voyages, and it is hoped that the advice and guidance contained in this book will continue to be a source of reference for you in the future. The charts that are included should be a practical guide in themselves, since many of them are marked with suggested patterns that can aid you in finding your way to objects efficiently.

Never forget the value of having an observing partner. If you do run into a brick wall between you and a deep sky object, it can be extremely helpful and encouraging to be able to ask someone next to you if they've found the object or if they can help you find it in your 'scope (Fig. 5.32). After reading a lot of information and reviewing star charts, sometimes hearing verbal advice from an observing buddy is just what is needed to solidify your understanding and help you to recall what you've read. Having a partner also gives you the opportunity to try out different optical equipment. Additionally, make it your goal to observe several times throughout the year. An annual observing session or two is simply not enough to get proficient at star pattern jumping through the constellations. Learn to be able to find several objects in the night sky no matter what season of the year you find yourself in.

Seeing galaxies of all types for the first time in their own telescopes causes aperture fever in many amateur astronomers. When the time is right, the increase in size can make a big difference if you enjoy deep sky observing. However, before you upgrade be sure to weigh the additional cost involved along with the considerations

Fig. 5.32 Working with an observing partner can be helpful when finding deep sky objects. (Image courtesy of the ESO)

of storage and portability. Also, be sure to thoroughly research the quality of any larger 'scope you are looking into. A well-made small telescope is far better than a large telescope that is poorly designed or constructed. No matter what size telescope you own, try to push the limits of what you can see. Never forget that published magnitudes are not the final word on whether or not you can see an object in your telescope or not. One well-known amateur astronomer commented in a discourse at ALCon (Astronomical League Convention) 2014 that every time he challenges his friend (who uses an 8-in. 'scope) to find a very dim object that should be out of range for his 'scope, his friend more often than not ends up seeing it.

Whatever are your favorite objects to observe, whether nebulae, galaxies, planets, or something else, strive to become an expert in that area. Repeated observations of objects that you are familiar with helps you to gain mastery at locating them and picking out the finer details. Just as important is locating new objects regularly. It keeps your mind fresh, open to discovering new things, and makes for more interesting observing sessions. This does not mean you should rush through a list just to check off several items. But if you just naturally pursue frequent and mean-ingful observing sessions that are well planned, the experience will come too. In other words, be *serious* about having *fun* when you observe, and the rest will likely take care of itself.

Helpful Things to Consider

Have I set a reasonable goal for myself in regards to the number of galaxies that I can locate in one observing session? Will I be searching for a very faint galaxy or one that is relatively bright?

For each galaxy I observed, was the type of galaxy readily apparent? After finding each target, did I make notes, including a sketch, so that it will be easier to find during my next observing session?

What unique features did I observe in each galaxy?

Did I stumble across any new galaxies nearby and did I consult my star atlas to identify them?

The number of galaxies that can be observed with a medium-size amateur telescope is virtually endless. Most of the galaxies described in this chapter can be found in small- to medium-size telescopes, but as was the case with nebulae and star clusters, more aperture usually results in a much more impressive view. Seeing photos of a galaxy beforehand can also be helpful, so that you have something to compare with what could potentially be seen in your telescope. Even more rewarding, however, is to simply see it for yourself and make a sketch at the eyepiece or soon after your session. This is one of the best ways to record those subtle details. To further enrich your viewing, you should learn as much as you can about the nature of the objects you observe. Knowing the story behind these awesome stellar furnaces and galactic clouds can elevate your understanding and how you approach visual astronomy. It will also illuminate and intensify the pure enjoyment that comes from observing the universe either alone or in the company of other people.

Chapter 6

Treat Yourself
to Dark Skies

It was not many years ago when dark skies were common all over the world. If we take a moment to reflect on the accomplishments of astronomers before the late twentieth century, we find that many of them were successful because of having easy access to dark skies. One wonders if the strides in our understanding would have occurred so quickly if light pollution had encroached on our skies earlier in history. Would there be a Messier catalog if Charles Messier lived in a time when the skies were brightly lit? Would William Herschel have left his career in music to observe if his sky was affected by sky glow? Would E. E. Barnard have gone on to become an astronomer if he was not able to see the Milky Way as a child? It is hard to say, but reflecting on those questions helps us re-think what we are doing to our current night skies. Let's briefly review some of the valuable contributions made to astronomy under conditions that were light-pollution free.

William Herschel is known as one of the greatest observational astronomers of all time. However, in the early part of his life he spent his time pursuing a career in music. By 1770 he had become a very capable musician, playing both the violin and oboe. His sister Caroline was also talented. She was a gifted singer, and after her brother William invited her to live with him in the town of Bath, England, in 1772, she had hoped for several years to develop her talent into a flourishing career.

However, in 1773 William's interest in astronomy really began to grow, and gradually his interest in it overshadowed his life as a musician, although he was still involved in music in order to make a living. Another pivotal year was 1778, during which Caroline had performed in several oratorios. After singing in a production of the *Messiah* in the spring of that year, she received an offer to sing in a concert that would have given her a real chance at a successful singing career. As a youth Caroline was prevented by her mother from acquiring any education that would allow her to obtain work outside of the home. Perhaps out of gratitude to William

© Springer International Publishing Switzerland 2015
D. A. Jenkins, *First Light and Beyond*, The Patrick Moore Practical Astronomy Series,
DOI 10.1007/978-3-319-18851-5_6

for inviting her to live in Bath, she turned down the offer to continue singing and decided to help her brother with his astronomical studies.

William Herschel's discovery of the planet Uranus in 1781 led to his receiving an annual salary from King George III. Now he could afford to give up work in music entirely and devote himself to astronomy. Caroline helped him observe night after night, recording objects he observed during his sweeps of the sky. While he would observe at the telescope, she was inside consulting charts for him such as Flamsteed's *Atlas*. Over time her own interest in observing grew, and she made many of her own discoveries. Between 1783 and 1787 she discovered two galaxies, and a total of eleven nebulae and star clusters. She went on to discover at least eight comets, and after her brother died in 1822, she completed a catalog containing 2,500 of the objects they had observed and recorded over the years. The Herschel catalog has since become the cornerstone of the *New General Catalog* (published in 1888) that we still use today.

Just a few decades later, E. E. Barnard, the young man from Nashville, raised without hardly any formal education, found his way into a simultaneous program of study and work as an astronomer at the Vanderbilt University Observatory in 1883. It was a slight cut in salary at first, but he believed it would be a once in a lifetime opportunity. We can all be glad that he did, since he went on to make many valuable contributions to astronomy. Just a few months prior to the offer from Vanderbilt University, local benefactors told him they wanted to build an observatory on his land that would have been named after him. He politely told his supporters that he could not accept their proposal. His real reason was that he knew if he accepted the gift of the observatory, he would forever feel obligated to stay there even if better opportunities presented themselves.

Barnard had come a long way from the day when he acquired his first telescope, which was only about a 1-in. refractor. Since then he graduated to the well-made 5-in. refractor, and he used it to make several discoveries that earned him the attention of Olin H. Landreth, then professor of Engineering at Vanderbilt University. By 1882 he had discovered several comets, and had been one of the first to observe at least six fragments of the Great Comet of 1882 after it broke apart. Barnard had an unmatched ability to observe the faintest of objects and see details where others could not. One example of this was in connection with Encke's Comet, which reappeared in 1885. Astronomer C.A. Young of Princeton University commented that he observed the comet with a 23-in. telescope but was not able to see it in a 9.5-in. telescope. However, Barnard was able to see Encke's Comet with his 5-in. refractor. Over the course of his life some of his significant accomplishments included the discovery of numerous comets, the fifth moon of Jupiter (Amalthea, discovered on September 9, 1892), dark nebulae, and Barnard's star (the second star closest to our Sun).

For each of those astronomers, the times in which they lived were essentially light-pollution free. There was no sky glow from a distant city interfering with their view of the first 30° above the horizon. There were no neighbors with bright flood lights shining into their backyards. No wonder that even with small telescopes, astronomers such as Charles Messier and E. E. Barnard were able to see so much.

Fig. 6.1 Views of the Milky Way are becoming rare for most of the world's population. (Image courtesy of ESO/Y. Beletsky)

In order to be as successful as them we need the same or near the same conditions. For most people that is just not possible, since the problem of light pollution has invaded even small communities that were once very isolated from bright city lights. So as amateur astronomers we have to make the best use of the skies we have access to. However this does not mean that our skies cannot be improved. In Chap. 2 we discussed some practical ways of improving the dark conditions of your local observing area. But there is more that can be done. Let's examine what is now being done about light pollution on an international scale (Fig. 6.1).

International Dark-Sky Association

The International Dark-Sky Association (IDA) is a non-profit organization that since 1988 has taken the lead in helping communities and local government become aware of the crisis that has so horribly impacted our night skies. Light pollution not only affects our ability to view the stars, but also seriously disrupts wildlife and the natural sleeping patterns of humans. A term often heard in connection with this is circadian rhythm. Circadian rhythms are physical changes that occur in humans and animals that follow a 24-h cycle. In both humans and animals, genes are responsible for directing our circadian rhythms.

The central nervous system controls these rhythms and allows for the healthy function of the brain and other essential bodily functions such as sleeping, maintaining healthy cells, hormone production in the glands, temperature control, and much more. It is the hypothalamus in the brain that regulates our resting during sleep, and our waking up for activity.

The natural cycle of light during the day and darkness at night is the main trigger that enables the body to operate properly. The nucleus of the hypothalamus is connected with fibers that come from the eye. That is why cycles of light and dark as seen by the eyes affect important functions in our bodies. Our circadian rhythms are upset during shift work, long periods of travel, or simply by a lack of sleep. Therefore the need for dark at night is important to health. The increasing flood of light at night not only disrupts humans but also does the same thing to wildlife. Much research has been done to demonstrate the bad effects of lighted buildings on birds, bats, insects, turtles, and sea life.

There is also a high annual cost of energy that is wasted. According to an IDA estimate, this cost is $2.2 billion each year in the United States. Because of the strenuous efforts of the IDA and many individuals who continue to champion its cause, many areas have adopted guidelines that protect our dark skies by restricting outdoor lighting. The result is a significant savings in the reduction of energy costs, and improvement to the health and well-being of humans and wildlife.

Since the IDA began its program in 2001, eight communities, eight reserves, and 19 parks had received recognition worldwide by 2014 as dark sky communities, reserves, or parks, with more on the way. In the United States there are 14 dark sky parks and six dark sky communities. Those six communities are: Borrego Springs, California; Sedona, Arizona; Flagstaff, Arizona; Dripping Springs, Texas; Homer Glen, Illinois; and Beverly Shores, Indiana. In the United Kingdom there are currently three dark sky reserves, two dark sky parks, and two dark sky communities (the Isle of Coll and the Isle of Sark). Also, in addition to these designations within the United Kingdom, there are also other sites (although not reserves) that have reasonably dark skies. These sites are listed on the Dark Sky Discovery website at http://www.darkskydiscovery.org.uk/dark-sky-discovery-sites/map.html.

In Europe there are three dark sky parks and three dark sky reserves recognized by the IDA. In other places Namibia, Africa, has been recognized as a dark sky reserve as well as the South Island of New Zealand. Other organizations involved in supporting these initiatives are Starlight, based in Europe, and also the Royal Astronomical Society of Canada (RASC). Since 1999 the RASC has designated 17 areas as meeting the requirements for Dark-Sky Preserves and two areas that have qualified as Urban Star Parks. Hopefully this is the beginning of a positive trend that will enable our dark skies to be preserved in many more places around the world.

How can a community become recognized by the IDA as a dark sky community? It begins with the people. People have to become concerned about light pollution and have the desire to mitigate it. Then a town's municipality must be made aware of their concerns, followed by a rigorous application process along with action taken to limit light pollution on a continual basis. Take the small town (population of about 620) of Beverly Shores, Indiana, as an example. The town is located on

Lake Michigan and sits on 2,300 acres of land, and about half of that is maintained by the National Park Service. Most people that live there do so because it is a quiet, peaceful community with access to an area rich in plant and wildlife as it borders the Indiana Dunes State Park. Although steps were taken in as early as 1983 to regulate outdoor lighting, it was in recent years that the town expanded its efforts so as to qualify under IDA guidelines in 2014. After property owners were fully informed as to what constitutes proper lighting, the town's residents raised the money to replace or retrofit all improper public streetlights.

Becoming recognized as a dark sky community goes beyond those initial steps. A community must have the ongoing support of entities such as its businesses, Chamber of Commerce, utilities, and home owner associations. A community must demonstrate its commitment by yearly distribution of dark sky awareness materials for residents and also by clearly demonstrating a marked improvement in light pollution control. A sky brightness measurement program must also be maintained by the community.

Another way that an individual or group can become involved is to form a local chapter of the IDA. When a member of the IDA forms a local chapter, it opens up access to much more support to help the promotion of dark skies locally. Many chapters have already been established, but often coverage of large regions (such as states) presents a lot of opportunity for more focused attention from newly formed local chapters. No special qualifications are required other than to be a zealous advocate for dark skies, and chapters must follow guidelines for non-profit organizations.

Some chapters specialize in working with local municipalities to effect change in lighting ordinances. Others choose to use their time and resources to educate the public. This is perhaps one of the most far-reaching ways of helping the IDA. When everyone is aware of how light pollution has impacted our environment, it can lead to individuals making changes on a personal level to lighting in their homes and businesses. In either case the important thing is to communicate in a positive way with others in the community and become partners in the mitigation of light pollution. The IDA provides a chapter handbook on their website for anyone interested in learning more about this important role.

Let's take some time to consider the comments of Dr. Connie Walker, who is an Associate Scientist and Senior Education Specialist at the NOAO (National Optical Astronomy Observatory in Tucson, Arizona) and director of the Globe at Night campaign. She is also on the Board of Directors of the International Dark-Sky Association. Being concerned about the growing loss of our pristine night skies, she is a strong advocate for initiatives that limit the effects of light pollution. During a recent interview, she offered her insight on this topic, which can help amateur astronomers understand the opportunity we all have to effect change in this area.

Author: *What first led you to become passionate about astronomy?*

Dr. Walker: Seeing some of the first pictures of Earth from the Moon's perspective and seeing the first man land on the Moon had a profound impact on me. The feeling of awe that comes from astronomy is overwhelming, the contemplation of the immensity and beauty of how things work. At times we can feel so insignificant, but yet we are not. For instance, the conditions for life may be common, but may be more arduous the more advanced life is. Earth could be special.

Author: *What specific areas in astronomy are important to you?*

Dr. Walker: Early in my career I used submillimeter-wave spectroscopy to study star formation in galaxies at different epochs of the universe. Now a large part of my focus is the preservation of our dark skies and public outreach. A lot of our culture has been influenced by dark, starry night skies, such as in the paintings of Van Gogh, the music of Holst's "The Planets," and the writings of Shakespeare, which feature several references to the skies including the planets, Moon, and the North Star.

With dark skies in cities disappearing slowly over the last 100 years, I fear that we are losing this as a means of inspiration for our culture. About one half (3.3 billion) of the world's population lives in cities. Also astounding is the fact that the majority of people living in those cities have not seen a dark, pristine sky. How do you convince people of what they have lost if they have never experienced it?

Author: *Is there anything new happening in your areas of interest?*

Dr. Walker: 2015 is the International Year of Light. One of IYL 2015's themes is called "Cosmic Light," which will focus on two areas. The first of the two areas is on dark skies awareness. We are providing quality lighting kits to promote awareness of light-pollution issues and responsible lighting solutions through a quality lighting education program and the Globe at Night campaign. The second part of our work will of course focus on astronomy. One way we do that is by providing an educational package along with a Galileoscope. We have created Galileoscope telescope kits that allow users to understand and see what Galileo first observed with his telescope more than 400 years ago. Our website is: light2015.org/Home/CosmicLight.

March 14, 2015, was (super) Pi ($\pi = 3.1415$) Day. The goal was to assemble people to take dark sky measurements in a flash mob format and submit them to Globe at Night. If the measurements were taken at 9:26 p.m. local time, then it will have been a very super Pi Day ($\pi = 3.1415926$). This joint, global effort was intended to gather data on the limiting magnitude at every location possible all over Earth, hence enhancing the efforts to monitor light-pollution efforts worldwide.

Author: *What can people do to curb the effects of light pollution in their cities and towns?*

Dr. Walker: Amateur astronomers can play a key role in curbing light pollution by educating the public about these issues so that more voices can be heard, and more positive results can be obtained in the future. Participate in the Globe At Night (globeatnight.org) campaign. So far, over 115,000 measurements from 115 countries have been taken of each area's limiting magnitude. Anyone can contribute local data by use of the report page on a smart phone in real time or later with a desktop computer and (for your "instrument") the naked eye, a sky quality meter, and/or through one of the phone apps available through the site (the Loss of the Night app or Dark Sky Meter app).

Why Observe Under Dark Skies

What difference does a dark sky make? A lot of features can be missed because of light pollution, and some objects are hidden altogether. The Astronomical League's Dark Sky Advocate program actually features several activities whereby the participant must observe and note how an object looks under heavy light pollution as compared with a dark sky. Many details are revealed in deep sky objects simply by way of having a dark, black background to provide a stark viewing contrast. A truly dark sky is like using a natural eyepiece filter. The pictures you see online or in your favorite astronomy magazine just do not do it justice. It is something that you really should experience for yourself. This is the hope of everyone who is trying to stop the spread of light pollution—to bring back this natural resource for everyone to be able to appreciate it.

In the pages to follow we will consider some additional objects that should interest you. Among these objects are some that you may not have considered possible to observe if you live in a light-polluted area. This does not mean that the objects are impossible to detect from moderately light-polluted skies. However, it does mean that these objects can be better appreciated from truly dark skies. If you make it a goal to travel to one of the many dark sites mentioned earlier or some other dark site, you will become a better visual observer. You will also come away with a newfound appreciation for the great lifetime pursuit of visual astronomy.

Such a trip would be a good time to bring an observing partner, but if you are going alone, be sure to always let someone know where you plan to be. It's also smart to take precautions, such as making sure your vehicle is in good working order (tires, car battery, headlights, etc.). Unlike Chaps. 4 and 5, where we covered specific types of objects one at a time, here we will cover a mixture of all kinds of deep sky objects as viewed in a 6-in. (152 mm), 10-in. (254 mm) or larger scope. Included are bright, easy objects such as clusters and asterisms, along with more challenging galaxies and nebulae that are much easier to track down in dark skies. After reading about them, get to a dark sky site and find out what they look like in *your* scope!

Additional Objects to Observe in Truly Dark Skies

The Crab Nebula, **M1** (NGC 1952), is an object that has a well-known history of observers that goes back for hundreds of years. It is the remains of a supernova explosion that was so bright it was visible in the daytime. In July of the year AD 1054 astronomers in Asia recorded the event. Pottery fragments of Native American drawings have been excavated in several states in the Southwest such as New Mexico. It is believed that they are records of the supernova from the same year. It was later observed by English astronomer John Bevis in 1731, and of course it was observed by Charles Messier. Messier first observed the nebula on

Fig. 6.2 The Crab Nebula (M1) in Taurus. (Image by the author)

August 28, 1758. The nebula is located about 6,500 light years from our Solar System and spans a distance of 10 light years. The gas outflow of oxygen, hydrogen, and sulfur from the explosion is still moving at a speed of 1,800 km per second. The faint central neutron star is a pulsar 10 km in diameter, pulsing every 33 ms (Fig. 6.2).

The Crab lies about 1° northwest of the third magnitude star Zeta (ζ) Tauri. It is clear why Messier included it in his catalog, as at first glance it could be mistaken for a large comet-like object. In a 6-in. telescope it appears hazy and indistinct. In a 10-in. or more it appears somewhat more composed and is much easier to recognize with the use of an O-III filter.

If you can get to more southern skies, the next two objects are spectacular to view in dark skies. The Centaurus constellation can be viewed at latitude 29° north or lower. Omega Centauri (**NGC 5139**) is one of the brightest globular clusters and even more impressive than M13 in Hercules. It is a bright naked-eye object, magnitude 4, and its ten million swarming stars fill the eyepiece. The cluster is 17,000 light years away and spans a distance of 50 light years! At the core of the cluster, stars are packed so tightly that it is quite possible that they are closer together than the distance from our Sun and the next nearest star 4.3 light years away. Current evidence seems to indicate a black hole is at its center. It is 7° east of the second magnitude star Gamma (γ) Centauri (Fig. 6.3).

The galaxy Centaurus A (**NGC 5128**) is a spectacular and unusual sight 11 million light years away. It probably resulted from the dynamic merger of two galaxies eons

Fig. 6.3 Omega Centauri NGC 5139. (Image courtesy of NOAO/AURA/NSF)

ago, which can still be seen today. This magnitude 6.8 galaxy can be seen easily with its dust lane in a 6-in. 'scope. In a 10-in. or larger the dust lane appears much thicker. The inner galaxy is a large elliptical that is encircled by a large spiral dust band whose shape undulates around its partner. To find this mesmerizing object, simply move 4.5° due north of Omega Centauri (Fig. 6.4).

A nice contrast to these objects is the open **Double Cluster** (NGC 884 and NGC 869) in Perseus, already discussed in detail in Chap. 4. However, with a transparent dark sky as its background it takes on a much more grandiose character. A very easy naked-eye target in truly dark skies, it hardly requires anything more than simply aiming your finderscope right onto the bright spot you see in Perseus. Figure 6.5 presents a wide field view of both parts of the Double Cluster together.

NGC 6210, the Turtle Nebula in Hercules, is magnitude 8.8, and although only 21″ (arc seconds) across appears bright. The nebula is about 6,500 light years away and was discovered in 1825 by German astronomer F. W. Struve. It is 4° northeast of Beta (β) Tauri, and 8° south of Zeta (ζ) Herculis, one of the stars that make up the Keystone asterism. It can be seen easily in a 6-in. 'scope or larger. For best views, use high magnification, which should reveal its resemblance to a turtle. You may observe a greenish hue in this planetary that further recalls its name (Fig. 6.6).

Fig. 6.4 Centaurus A (NGC 5128). (Image courtesy of ESO)

In Hydra the 7.3 magnitude globular cluster **M68** (NGC 4590) is a beautiful sight. The 10-billion-year-old cluster was discovered by Charles Messier on April 9, 1780. It is about 35,000 light years away and spans 140 light years across. The cluster is very easy to locate just 3.5° south of Beta (β) Corvi. Also in the constellation of Hydra is the Ghost of Jupiter (**NGC 3242**), less than 2° south of Mu (μ) Hydri. This planetary nebula contains a bright inner shell, surrounded by a fainter outer layer. Astronomers have detected strong stellar winds and X-ray emissions. It is only about 3,260 light years away and is just 4,000 years old. In a 6-in. 'scope its small orb can be seen. In a 10-in. or larger 'scope it is very easy to see and has a blue-green color, and some report seeing color in smaller 'scopes. It responds well to higher magnification. Another planetary that shows up easily under dark skies is **NGC 4361**, in Corvus. You will find it located 2.5° southeast of Gamma (γ) Corvi, forming a triangle along with Delta (δ) Corvi. It is about 2,600 light years away

Fig. 6.5 Wide field view of the Perseus Double Cluster. (Image by the author)

and at about magnitude 11 is challenging in small 'scopes. It has a size of about 2′ (arc minutes) and is very easy to locate with a 10-in. or larger 'scope (Fig. 6.7).

We now turn our attention to a few galaxies in the constellation Boötes the Herdsman. Boötes is home to the bright star Arcturus, the first star believed to be observed in daytime in the year 1635 by French astronomer Jean-Baptiste Morin. Although it contains none of the Messier objects, it is home to 493 NGC and IC objects. **IC 1029** and **NGC 5673** is an interesting pair at magnitude 11.3 and 12.1 respectively. They may be binary galaxies that orbit each other. These two spiral galaxies are about 78 million light years away and are also considered to be part of a physical group that includes at least two other galaxies (NGC 5676 and NGC 5660). In a 10-in. 'scope or larger, they appear so similar that they resemble a pair of glowing eyes in the dark.

The four galaxies are located 11° northwest of 3.5 magnitude Beta (β) Boötis and 2° southeast of the fourth magnitude Theta (θ) Boötis. Figures 6.8 and 6.9 show both the broad and enlarged areas to help you pinpoint their exact location.

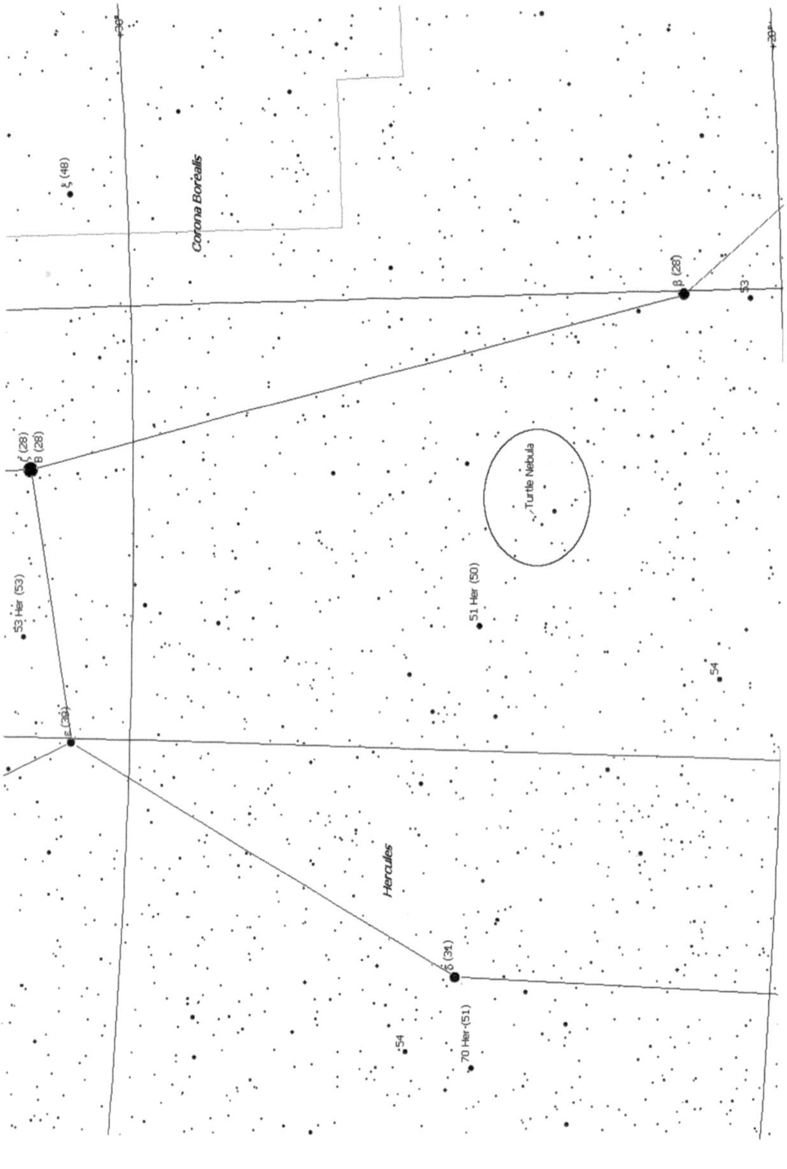

Fig. 6.6 Area near the Turtle Nebula (NGC 6210). (Star chart courtesy of *Sky Tools 3* at: www.skyhound.com)

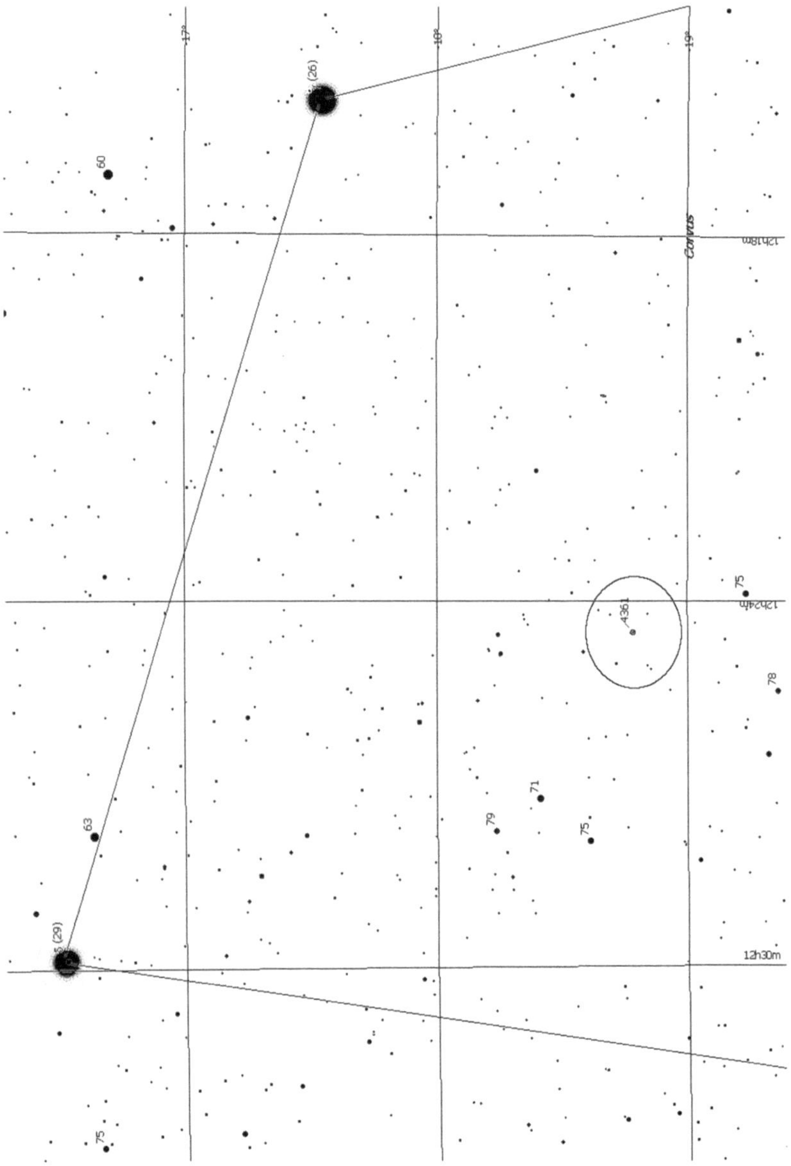

Fig. 6.7 Area near NGC 4361 in Corvus. (Star chart courtesy of *Sky Tools 3* at: www.skyhound.com)

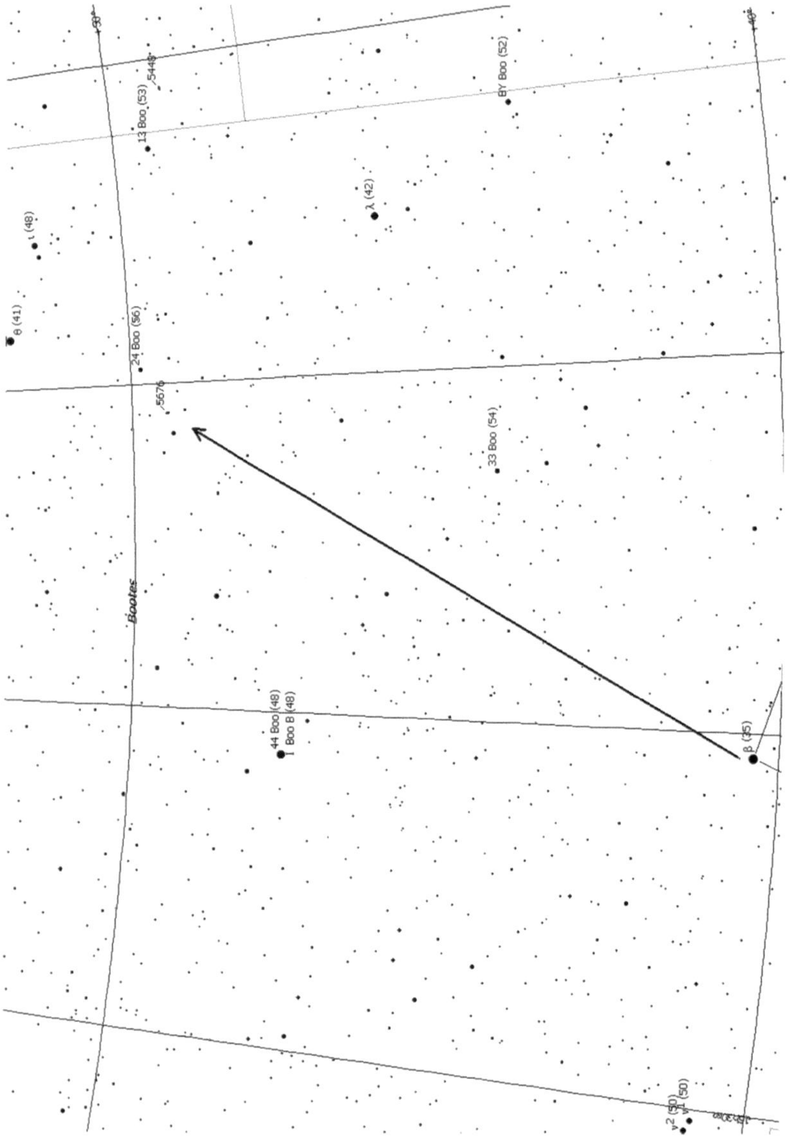

Fig. 6.8 Broad area near NGC 5676. (Star chart courtesy of *Sky Tools 3* at: www.skyhound.com.)

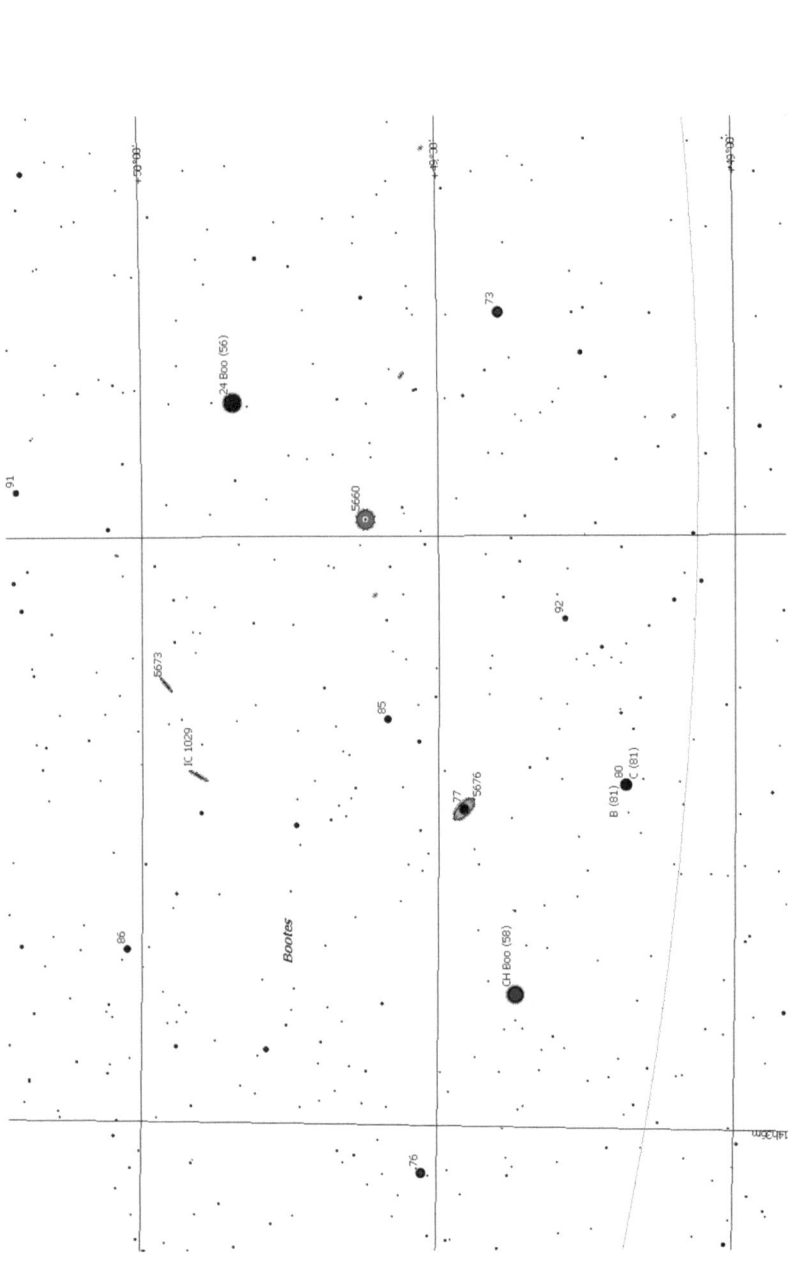

Fig. 6.9 Enlarged area near the four galaxies IC 1029, NGC 5673, NGC 5676, and NGC 5660. (Star chart courtesy of *Sky Tools 3* at: www.skyhound.com)

Fig. 6.10 Spiral galaxy NGC 5676 in Boötes. (Image by the author)

NGC 5676 is magnitude 11.2, and its companion **NGC 5660** is magnitude 11.9. Both are just 30′ to the south of the previously mentioned pair, but are wider apart from each other. Astronomers believe the evidence supports a high amount of dark matter within NGC 5676. It appears rather large and bright under dark skies with a 10-in. 'scope or larger. However, NGC 5660 has a much lower surface brightness and appears ghostly even in a 12.5-in. 'scope (317 mm). This pair is about 108 million light years distant (Fig. 6.10).

Located 3° west-northwest of Algol (second magnitude star Beta [β] Persei) is the spiral galaxy **NGC 1122**, an object also included in the Herschel catalog. At magnitude 12.1, it is quite a challenge unless you are in truly dark skies. Although it has not been the subject of a lot of study, NGC 1122 is a very beautiful galaxy about 160 million light years away, and has a large diameter of 82,000 light years (Fig. 6.11). This spiral is quite easy to see in a 10-in. or larger telescope. By way of contrast the bright, easy to find open star cluster **M34** sits just 2° west of NGC 1122. The 200-million-year-old open star cluster is just 1,600 light years away.

After trying a time or two from moderately light-polluted skies, many amateur astronomers have given up trying to see the dark nebula cataloged by E. E. Barnard in 1913. It is about 2 light years wide and 1,600 light years away. The **Horsehead Nebula** (Barnard 33 and IC 434, the bright nebula behind it) is normally a very illusive target. With dark skies, however, you can see this fascinating sight. It is a difficult object in a 6-in. 'scope but possible (Fig. 6.12).

Fig. 6.11 Spiral galaxy NGC 1122 in Perseus. (Image by the author)

Fig. 6.12 The Horsehead Nebula in Orion. (Image by the author)

In a 12.5-in. (317 mm) 'scope under very dark skies and the best observing conditions, you can see the shape of the Horsehead clearly visible amidst the brighter surrounding nebula. A filter is not even needed to see it, but you may see some improvement with a hydrogen-beta filter. In a transparent sky during winter,

Fig. 6.13 NGC 1977, 1981. (Image by the author)

it gives the appearance of being three-dimensional, as if its familiar shape had been carved out by celestial forces. Under the right conditions, this nebula is easy to find just south of the easternmost star in Orion's Belt.

While you are exploring all of the other nebulae in the vicinity (such as IC 435, NGC 2023, and NGC 2024) be sure to check the Orion Nebula again. If you have not been able to see the four stars of the Trapezium, you will certainly be able to see them under a pristine dark sky, even with a small telescope. Just 2.5° southwest of the Horsehead Nebula is another attractive nebula in Orion. **NGC 1977** and its related open star cluster **NGC 1981** are often overlooked when viewing the Orion Nebula just to the south (Fig. 6.13).

About 30 % of very distant galaxies are interacting galaxies. Another good example of this that can be seen in amateur telescopes is the pair known as the Antennae Galaxies, **NGC 4038** and **NGC 4039**. Situated 70 million light years away in the Corvus constellation, this pair is rich in star birth, much of it unseen in visible light but revealed by submillimeter-wavelength observations made by 12 antennas from ALMA (Atacama Large Millimeter/submillimeter Array in Chile, which will consist of 66 antennas upon completion). At magnitude 12.0 they are well within reach of amateur telescopes. A 6-in. 'scope will reveal the billowy mass of galactic matter, but in a 12.5-in. 'scope the shape becomes very clear, resembling a heart shape. The long wisps that give them their distinctive name, are difficult to see in anything but a very large 'scope. To find this pair, look 3.5° west-southwest of Gamma (γ) Corvi. Prominent in that area is the fifth magnitude variable star TY Corvi, less than 1° south of the Antennae. At least seven other galaxies under magnitude 13.0 are strewn about within a 4° radius. How many of them can you locate and identify using your own star atlas or the accompanying chart? (Fig. 6.14)

Now let's turn to an easy to find object that has a special quality. It is a deep sky object that presents the best of both worlds. A beautiful open cluster is combined

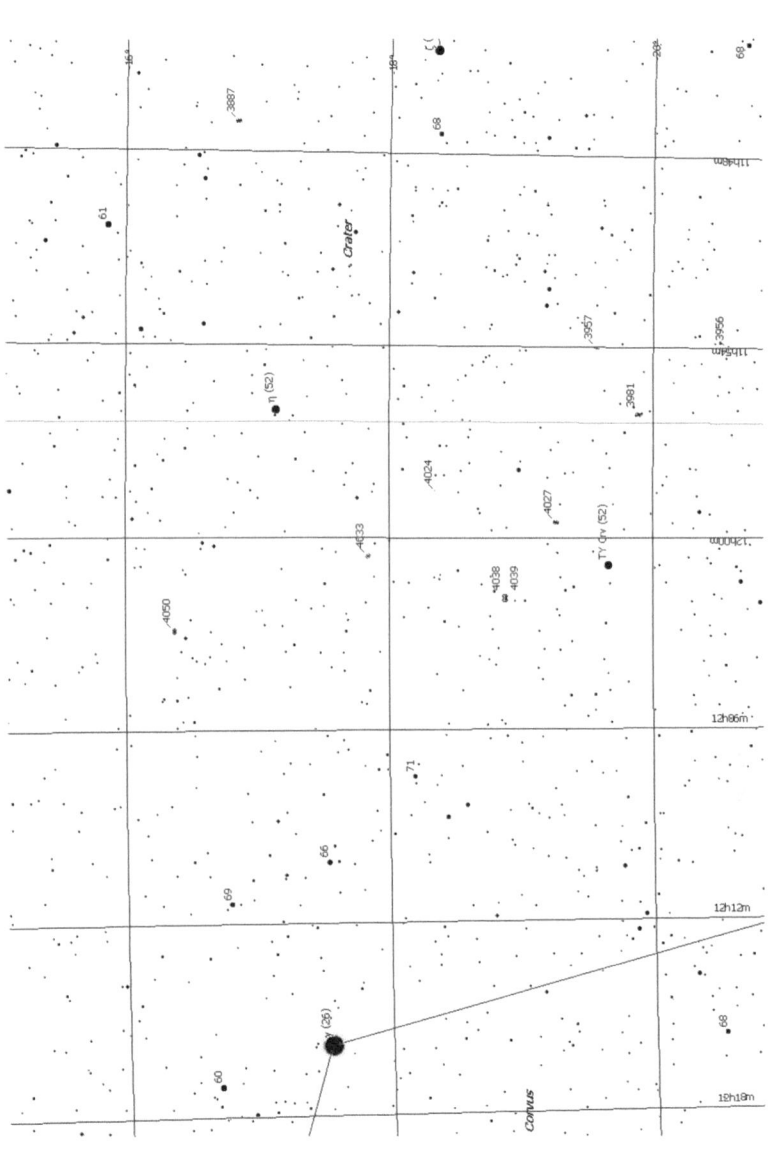

Fig. 6.14 Area near the Antennae Galaxies. Several other galaxies scattered in the region are within range of amateur telescopes. (Star chart courtesy of *Sky Tools 3* at: www.skyhound.com)

with a nebula. It is **NGC 281,** otherwise known as the Pacman Nebula, and the cluster is **IC 1590**. The cluster spans a distance of about 21 light years and is rich with new star formation. The Pacman Nebula is about 9,500 light years away. Astronomers have found that NGC 281 sits about 1,000 light years above the Perseus arm of our galactic plane and consists of a large Hydrogen II region.

Some of the research involving this nebula focused on the analysis of water masers in it. These are H_2O molecules that have been excited by radiation to become amplified and thus emit more radiation in what has been compared to a microwave laser. It is likely that the gas in this nebula was blown out of our galactic plane in a superbubble, due to several explosions of supernovae [8].

You will find the Pacman Nebula 1.5° due east of Alpha (α) Cassiopeiae. After comparing a low power view with a medium power view, you may prefer very low power. In a 6-in. 'scope the view is just perfect at 30×. No filter is required to see this intriguing object. It is like viewing the M13 cluster but with lots of nebulosity! (Fig. 6.15).

The Butterfly Nebula, **NGC 6302** (also called the Bug Nebula), at magnitude 9.6, presents another great show in a 6-in. 'scope. It has a beautiful structure that has been developing over the last 2,200 years. The nebula is 2.7 light years in size and is located 3,800 light years from us in the Scorpius constellation. The wings of this butterfly are streams of gas in excess of 36,000° Fahrenheit, flowing at speeds of about 600,000 miles per hour (965,606 km per hour). The surface of the central star is estimated to be about 400,000° Fahrenheit. To locate the Butterfly Nebula, first find the bright star Lambda (λ) Sagittarii. Then move 4° due west, following a trail of several sixth and seventh magnitude stars along the way. An eighth magnitude star (HD 156156) lies just 47′ before (east of) arriving at the Butterfly (Fig. 6.16).

Open cluster **M23** (NGC 6494) is a beautiful object in dark skies. It is a bright magnitude 5.5 object about 2,000 light years away. Its 150 stars are 220–300 million years old, spread across a distance of about 16 million light years. An easy cluster for you to locate in binoculars, travel about 9.5° northwest of third magnitude Kaus Borealis (Lambda [λ] Sagittarii) in Sagittarius. Nearby is a nice planetary nebula, **NGC 6445**, just 2° southwest. At a distance of 4,600 light years, NGC 6445 sits further away from us than M23. It is magnitude 11.2 and easy to spot under low to medium power in a 6-in. 'scope. Using a low power eyepiece, you will see a 7.5 magnitude star that is 600 light years away in the same field of view as this planetary nebula that is so much farther from us.

Within just 22′ to the south of NGC 6445 is a challenging little globular cluster, **NGC 6440.** The age of this cluster is approximately 10 billion years. At magnitude 9.3 it is not particularly bright for a globular, and it is only 4.4′ in diameter. This cluster is 23,000 light years distant and is very small compared to the diameter of the globular cluster M92 in Hercules, which is 14′ in diameter. However, part of the allure of observing globular cluster NGC 6440 is its tiny size. Although small in a 6-in. 'scope or larger it will seem to pop into view (Fig. 6.17).

The **Veil Nebula** (NGC 6992 and NGC 6960) is an absolutely stunning sight. This is especially the case with fast f/5 or f/6 refractors that provide a very wide field of view. The Veil Nebula is made up of the remains of a supernova explosion

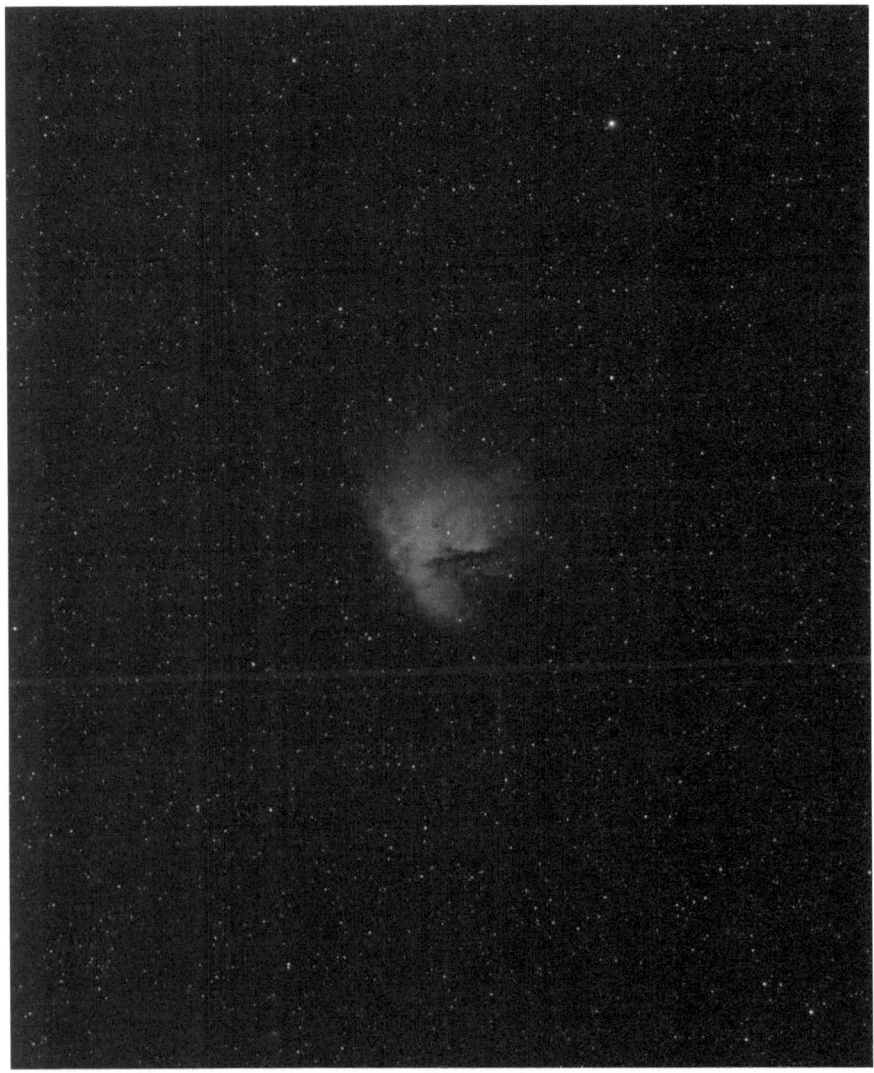

Fig. 6.15 The Pacman Nebula (NGC 281). (Image by the author)

that occurred about 10,000 years ago. Situated about 1,500 light years from us in the constellation Cygnus, it is sometimes referred to as the Cygnus Loop. Small telescopes will show this nebula easily. Under dark skies no filter is needed to enhance the view, especially for large 'scopes. As always, however, feel free to use any filters you own or can perhaps borrow, to see if the view is enhanced. The western portion of this nebula contains the beautiful star 52 Cygni, which is close to us at a distance of only 48 light years. It is superimposed on our view of the

Fig. 6.16 The Butterfly Nebula (NGC 6302). (Image by the author)

western part of the Veil (NGC 6960), which is 30 times further away. You will find it 14° south of the bright star Deneb (Alpha [α] Cygni). From Deneb, move south 11° to Epsilon (ε) Cygni. The Veil Nebula is 3° south of that 2.5 magnitude star (Figs. 6.18 and 6.19).

In the constellation of Monoceros we find several deep sky objects that you will want to put on your observing list. The first object we'll consider is **NGC 2316**, a small comet or cone-shaped nebula. NGC 2316 is actually a young cluster of stars embedded in a cloud of Hydrogen II that is powered by a hot blue star, where new star birth continues to occur. The cluster is 3 million years old and resides approximately 3,500 light years from us. It is not hard to find under dark skies with a 6-in. 'scope. Using an O-III filter on this object offers little improvement. However, in a 10-in. or larger 'scope it does appear much brighter.

Just 1° southeast of NGC 2316 is the open cluster **M50** (NGC 2323), first observed by Messier on April 5, 1772, but initially discovered in 1711 by French astronomer Giovanni Domenico Cassini (1625–1712). M50 is about 3,300 light years distant, contains over 2,000 member stars, and is about 130 million years old. At sixth magnitude it is an easy target for the naked eye and binoculars. To locate

Fig. 6.17 Area near the open cluster M23 in Sagittarius. Note NGC 6445 and NGC 6440 also. (Star chart courtesy of *Sky Tools 3* at: www.skyhound.com)

Fig. 6.18 The Veil Nebula (NGC 6960). (Image by the author.)

both of these objects begin with Sirius. Then move exactly 5° north-northeast to fourth magnitude Theta (θ) Canis Majoris. Just 4° more to the north-northeast is M50. From there the open cluster and nebula NGC 2316 is less than 1° to the north-west of M50.

The last object to consider in Monoceros is the exciting **Rosette Nebula (NGC 2237–2239)**, which surrounds an open star cluster (NGC 2244) that is the easiest part of this area to locate. Although it appears as one object in photos, it was cataloged as three separate NGC numbers. Modest 'scopes easily reveal the cluster, but even in dark skies the nebula itself is a challenge. An O-III filter will make quite a difference in being able to see it, as will larger apertures. Do not expect to see the characteristic red color present in many photos. However, the nebula is just as beautiful to observe even in the grayish tones interpreted by our eyes. In total there are about 67 NGC and IC objects in Monoceros, so before you leave be sure to take your time and explore the entire constellation's celestial wonders (Fig. 6.20).

Now let's return to Orion to examine an often overlooked nebula, **M78** (NGC 2068). M78 is a reflection nebula located 1,300 light years away and is a part of the giant molecular cloud, Orion B. Although much smaller than M42 (the Orion Nebula), M78 is bright and easy to find. It was first observed by Pierre Méchain in the beginning of 1780, and then Messier observed it on December 17, 1780. It is just one part of several nebulae that have been cataloged separately. The other parts adjacent to or

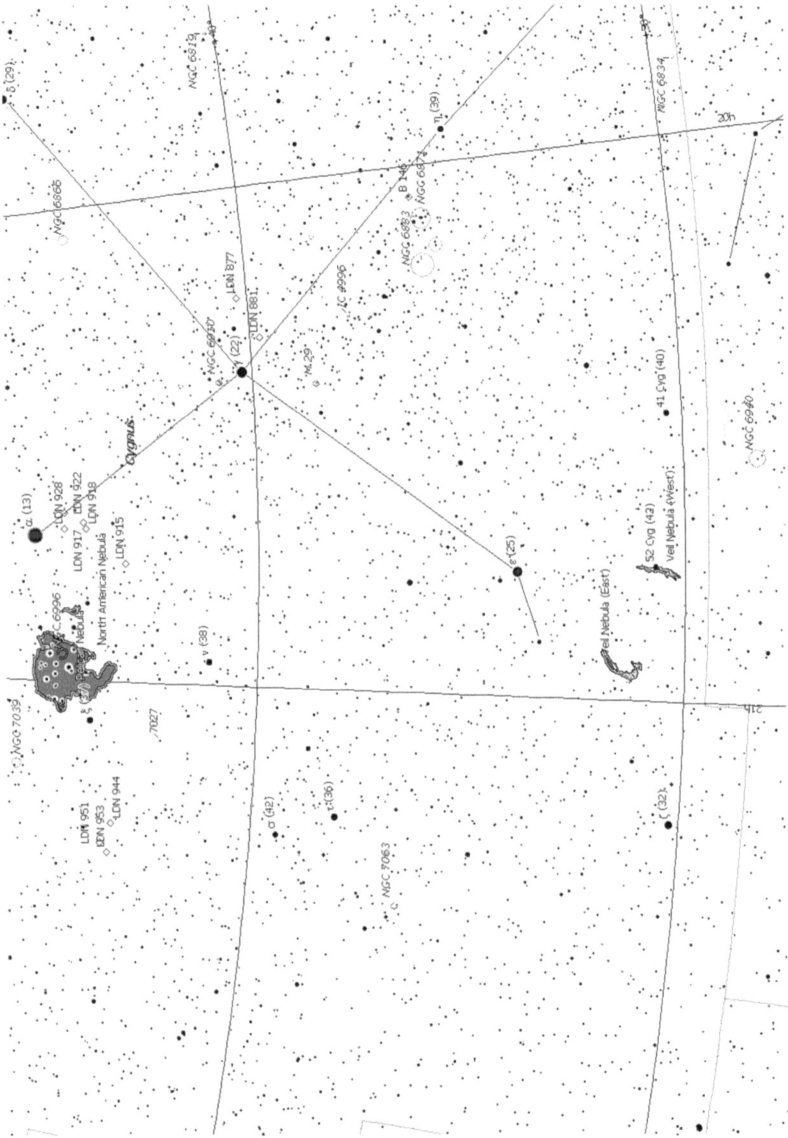

Fig. 6.19 Area showing the location of the Veil Nebula. (Star chart courtesy of *Sky Tools 3* at www.skyhound.com)

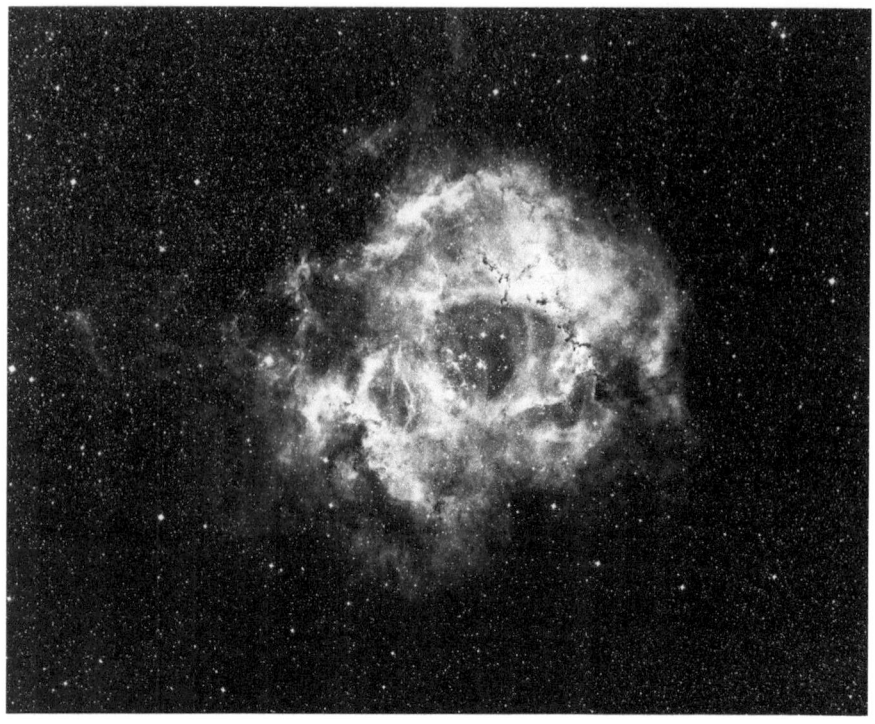

Fig. 6.20 The Rosette Nebula. (Image courtesy of NOAO/AURA/NSF)

near M78 are NGC 2071, NGC 2067, and NGC 2064. Additionally, amateur astronomer Jay McNeil imaged a portion of the nebula previously unidentified just south of the other well-known portions. McNeil's Nebula was not captured in very many photos prior to his imaging it. This had led researchers to believe that this sudden flare-up was due to an outburst caused by increasing accretion within a circumstellar disk. You can find this entire area of nebulosity by looking just 2.5° north-northeast of Zeta (ζ) Orionis (the easternmost star in Orion's Belt). Figure 6.21 contains a chart showing an enlarged view of the area.

   NGC 7789 in Cassiopeia is an excellent open cluster to observe under dark, transparent skies. Although cataloged by William Herschel in 1787, it was actually discovered by his sister Caroline Herschel on October 30, 1783. The 600 or so stars of this loose but very large cluster resolve very well in small telescopes and look particularly stunning in a 6-in. or larger 'scope. It is easy to become captivated by this beautiful group while trying to take it all in. NGC 7789 contains an abundance of different types of stars, including main sequence stars, helium burning stars, and blue stragglers. The age of this cluster is about 1.6 billion years, and it is located about 9,000 light years away. You will find this cluster 3° southwest of Beta (β) Cassiopeiae.

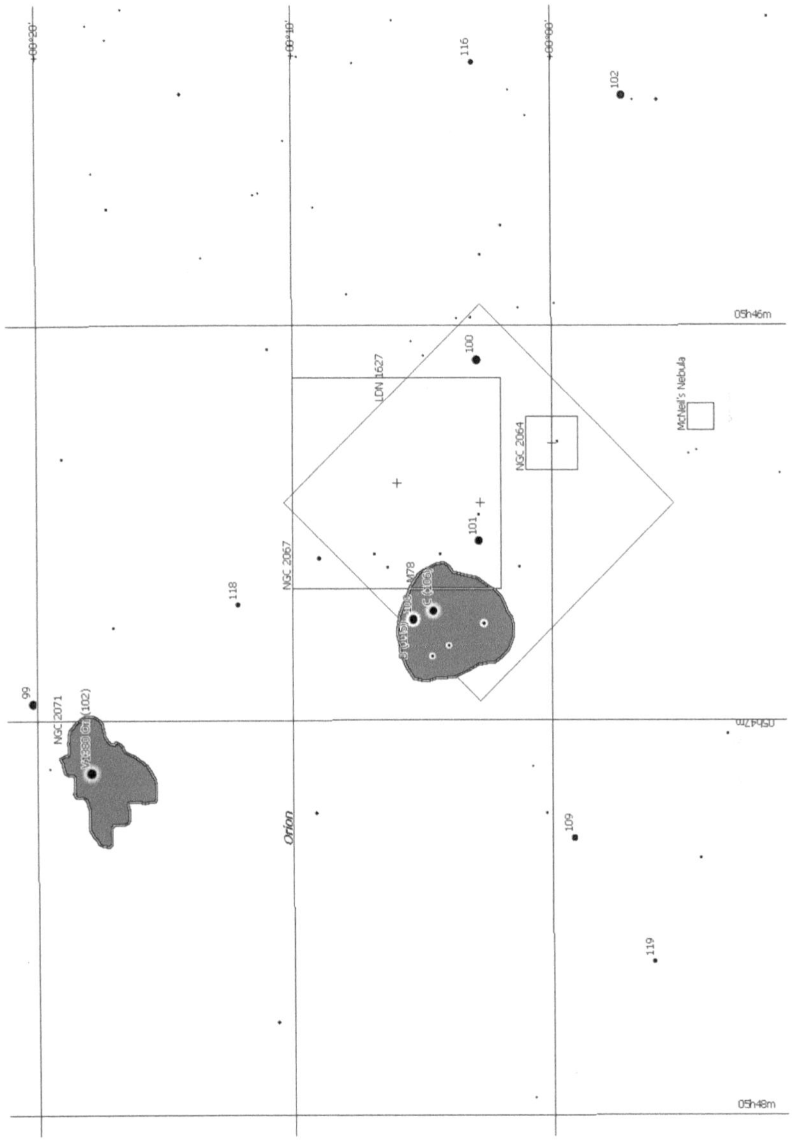

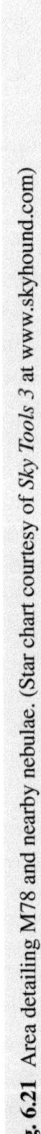

Fig. 6.21 Area detailing M78 and nearby nebulae. (Star chart courtesy of *Sky Tools 3* at www.skyhound.com)

Also in Cassiopeia is **NGC 659**, another beautiful open cluster that is 11,400 light years away. Its 200 member stars are about 22 million years old. This cluster is beautiful in a 6-in. 'scope. Notably near its center is a pretty little yellow-orange star. Like the older and larger NGC 7789, NGC 659 was also discovered by Caroline Herschel, on September 27, 1783. To locate it find third magnitude Delta (δ) Cassiopeiae, move 1° northeast to M103. Then from M103 move east just 80′ to NGC 659.

We now return to Pegasus, where several galaxies will capture our attention. The constellation of Pegasus ranks number 7 out of all the constellations in regards to size. It covers about 1,120 square degrees. Pegasus has only one Messier object (M15) but contains 433 NGC and IC objects within its borders.

The first galaxy to examine is **NGC 7457,** which is considered a lenticular galaxy because of its lens-like appearance. Lenticular galaxies also have a disk of stars and a bulge but do not show spiral arms characteristic of spiral galaxies. NGC 7457 is an interesting object because astronomers have found that it has a counter-rotating core, spinning in the opposite direction from the galaxy's disk. The stellar population in the bulge is approximately 7 billion years, whereas the stars in the nucleus are only about 3 billion years old.

Also, like our own galaxy and other large galaxies, NGC 7457 has been found to contain at least 200 globular clusters. The lenticular galaxy is a distance of about 40 million light years and is receding away from us at a speed of at least 790 km per second. Its magnitude is 11.2, and its apparent size is 4.3′ × 2.3′. Visually perceptible in a 6-in. 'scope, it appears large in a 12.5-in. telescope or larger.

To view this lenticular, first locate 2.5 magnitude Scheat (Beta [β] Pegasi) in the Great Square of Pegasus. NGC 7457 is just 2° north of β Pegasi. Use the accompanying star chart to help you get there efficiently. Note the patterns of some of the other nearby stars to aid you in making sure you end up in the right spot (Fig. 6.22).

NGC 7625 is a spiral galaxy (non-barred) that is positioned face-on to our point of view. The galaxy was also classified as a peculiar galaxy by Halton Arp and is also designated as Arp 212. It is about 77 million light years distant and is speeding away from us at 1,864 km per second. Although faint at magnitude 12.1 it can be seen in a 6-in. 'scope under good conditions. A magnitude 6.7 star 220 light years away is just 7′ east of this galaxy, which makes for a nice pairing within the same field of view. In a 12.5-in. 'scope the galaxy appears bright and large, perhaps in part due to its orientation. In very large amateur 'scopes, observers can try to see two dust lanes in the galaxy that resemble tubes. Begin with Alpha (α) Pegasi to find it. Then move just over 4° northeast to this peculiar galaxy. On the way there are two interesting groups of stars, the first looks like a lower case letter "t." The next group looks like a cascading waterfall of stars. Figure 6.23 should help you pinpoint these markers.

The two spiral galaxies **NGC 7332** and **NGC 7339** in Pegasus are also interesting. Evidence points to these two being an interacting galaxy pair. NGC 7332 has a large bar inside of it almost 10,000 light years long, and its existence seems to indicate that it had a recent interaction with NGC 7339. Inside of the central region of NGC 7332 are counter-rotating stars and gas. Its stars have a uniform average age of 6 billion years old throughout its diameter of 59,000 light years. NGC 7332

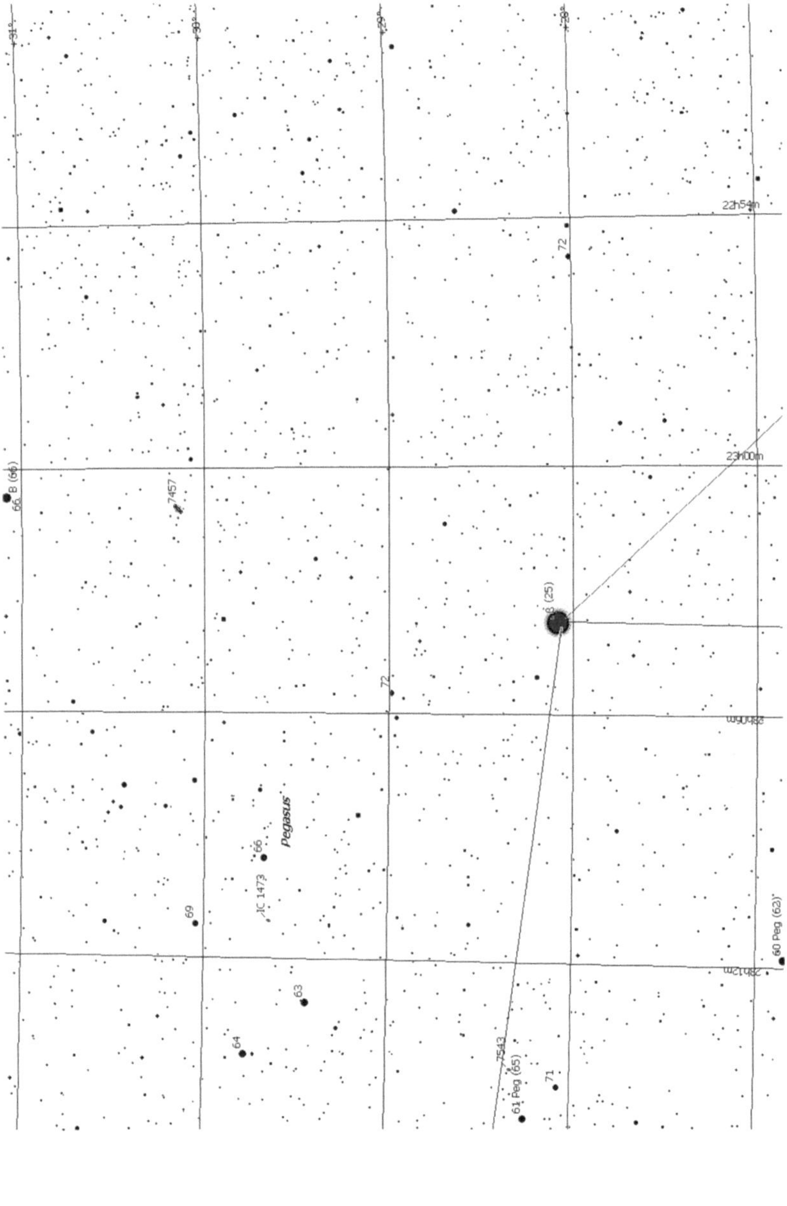

Fig. 6.22 Enlarged area showing NGC 7457 near Beta (β) Pegasi. (Star chart courtesy of *Sky Tools 3* at www.skyhound.com)

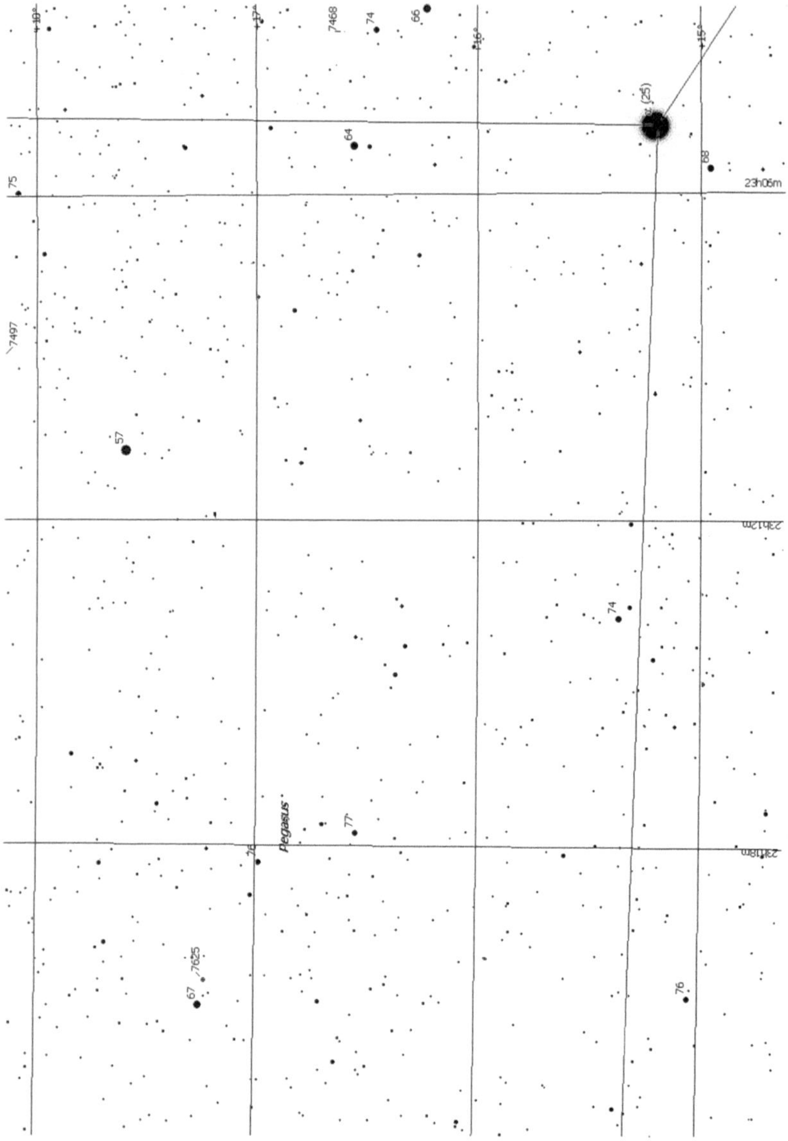

Fig. 6.23 The area near peculiar galaxy Arp 212 (NGC 7625) in Pegasus. (Star chart courtesy of *Sky Tools 3* at www.skyhound.com)

is 75 million light years distant, and NGC 7339 is slightly closer to us at 73 million light years distant, and has the same diameter as its partner. A distance of about 114,000 light years separates the two, and both are receding away from us at about 1,200 km per second. Visually, this galaxy pair is a real treat to observe. NGC 7332 is magnitude 11.1 and NGC 7339 is magnitude 12.2, well within reach of medium-size 'scopes. The dimmer galaxy appears thin, and the twin brighter galaxy has a much more bulged effect. Just south of NGC 7332 is an 11th magnitude star. The pair is about 2° west of the fourth magnitude star Lambda (λ) Pegasi. Both galaxies sit between two seventh magnitude stars, running north to south, on the western side of a triangle of stars (Fig. 6.24).

Now let's take a slight detour to a constellation in the southern summer sky. We arrive at Leo to observe an interacting galaxy pair about 67 million light years away. **NGC 3226** is an elliptical galaxy (magnitude 11.3) that sits on the northern edge of spiral galaxy **NGC 3227** (magnitude 10.3). Both were included in William Herschel's catalog and are also a peculiar galaxy pair designated as Arp 94. Imaging of the two galaxies in various wavelengths such as the ultraviolet have revealed a large cloud of neutral hydrogen (HI) nearby that is about 29,000 light years long and 20,000 light years wide. The cloud lies in front of NGC 3227 as compared to our line of sight. Astronomers have found heavy star birth in the southern portion of this HI cloud indicated by the existence of blue knots [9]. It is possible that this cloud is in the early stages of becoming a tidal dwarf galaxy, using material that it is taking from NGC 3227. The stars within the cloud are less than 100 million years old. NGC 3226 is also known to be a Seyfert galaxy.

Visually this interacting pair is easy to observe in a 6-in. or larger 'scope. First find Gamma (λ) Leonis in the mane of the Lion, the bright second magnitude star named Algieba. You will see the pair less than 1° to the east of Algieba. Once you edge Algieba just out of the field of view, removing its glare, both NGC 3226 and 3227 should snap into view. The elliptical shape of NGC 3226 can be clearly distinguished from the much larger spiral to its south (Fig. 6.25).

Our next stop is near the star Theta (θ) Leonis. **NGC 3596** is a large face-on spiral galaxy at least 50 million light years distant. It is somewhat challenging in a 6-in. 'scope but easy in a 10-in. or larger, appearing bright and well defined. To find it, look 40′ south of 3.4 magnitude θ Leonis. For another easy galaxy to locate, continue 4° southeast from NGC 3596 and past the Leo Triplet discussed previously. There you will find magnitude 12.0 **NGC 3666**, paired with a beautiful sixth magnitude yellow star about 500 light years away. NGC 3666 is a spiral galaxy about 43 million light years distant. The fourth magnitude star Iota (ι) Leonis is less than 1° to its south. Review Fig. 5.17 for a detailed look at this area containing several galaxies.

For a change of pace, let's consider an asterism commonly known as the Coathanger, or **Collinder 399**. This familiar-looking group of stars is in the Vulpecula constellation, just 2° west of fifth magnitude 9 Vulpeculae. There are many asterisms to be observed and a dark sky provides a great setting in which to do so with much better naked-eye visibility. The Astronomical League offers an observing program called the Asterisms Observing Program that includes a total of 108 different ones.

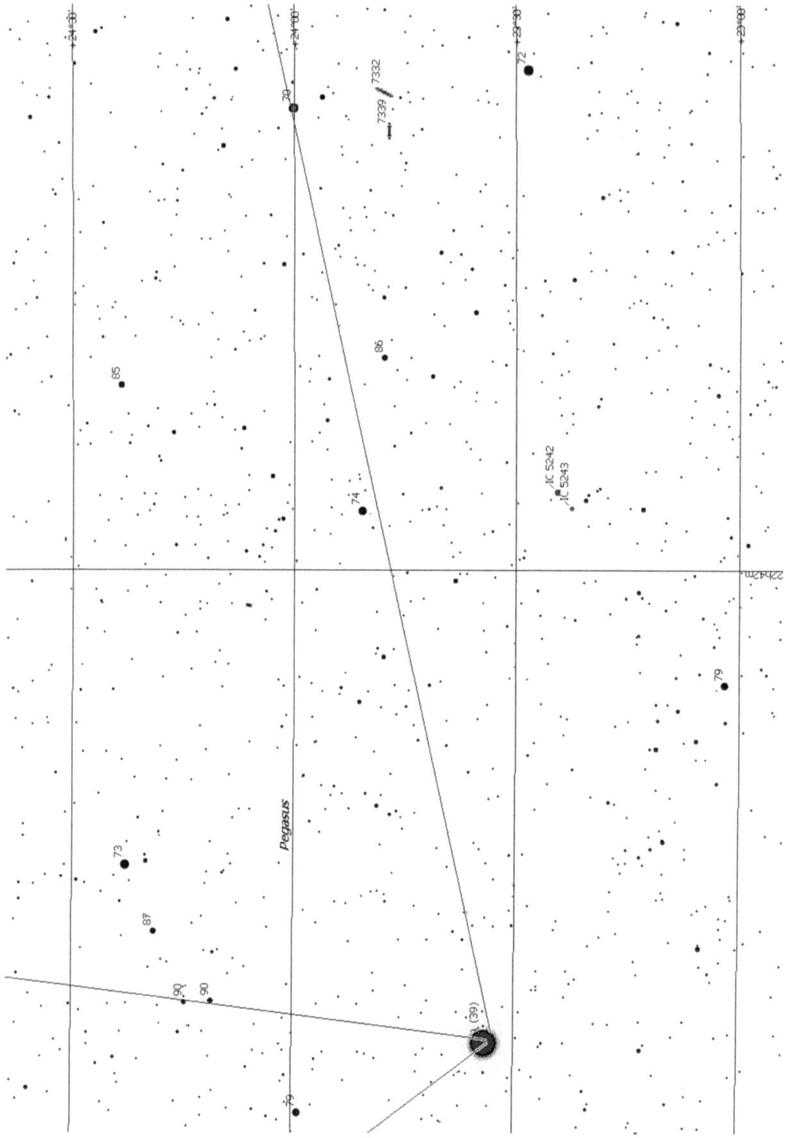

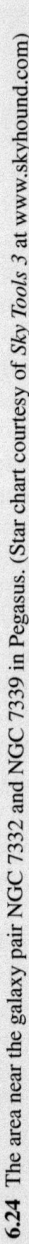

Fig. 6.24 The area near the galaxy pair NGC 7332 and NGC 7339 in Pegasus. (Star chart courtesy of *Sky Tools 3* at www.skyhound.com)

Fig. 6.25 The interacting galaxy pair NGC 3226 and NGC 3227 in Leo. (Image by the author)

Some are visible with only the naked eye, others in binoculars, and still others are visible only in a telescope. The stars seem to entertain us with the various shapes that are formed, and completing this program is a good way to familiarize yourself with the constellations and have a lot of fun at the same time.

We'll conclude this tour of objects with a recommendation that you visit the Auriga constellation to observe two dynamic nebulae, **IC 405** (The Flaming Star Nebula) and **IC 410**. IC 410 is the easier of the two nebulae to observe, but both show up clearly with an O-III filter and are only separated by about 1.5°. First locate Iota (ι) Aurigae, then move about 5° west to find IC 410. IC 405 is to the northwest of it. Both are great to view amid an area rich with many other objects to observe (Fig. 6.26a, b).

Planning a Trip to an Observatory

When it launched on April 24, 1990, the Hubble Space Telescope carried with it all of our hope and dreams. However, it wouldn't be until after initial repairs, completed on January 13, 1994, when its new corrected images were released, that it would become everything that we could have imagined. A telescope in space, free of Earth's atmosphere and so beautiful, floating out there doing the seemingly impossible. This telescope has brought the universe into our homes and has made our

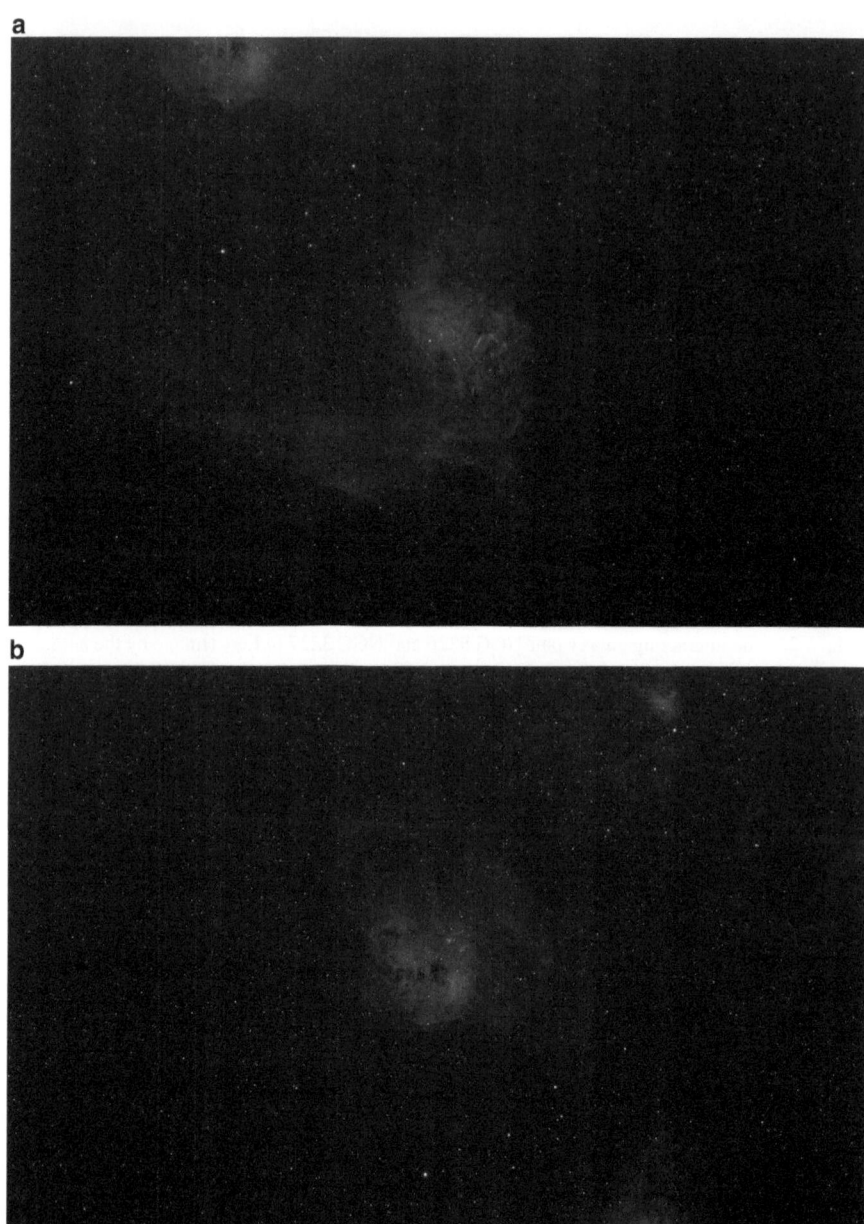

Fig. 6.26 (**a**) IC 405 in Auriga. (Image by the author.) (**b**) IC 410 in Auriga. (Image by the author)

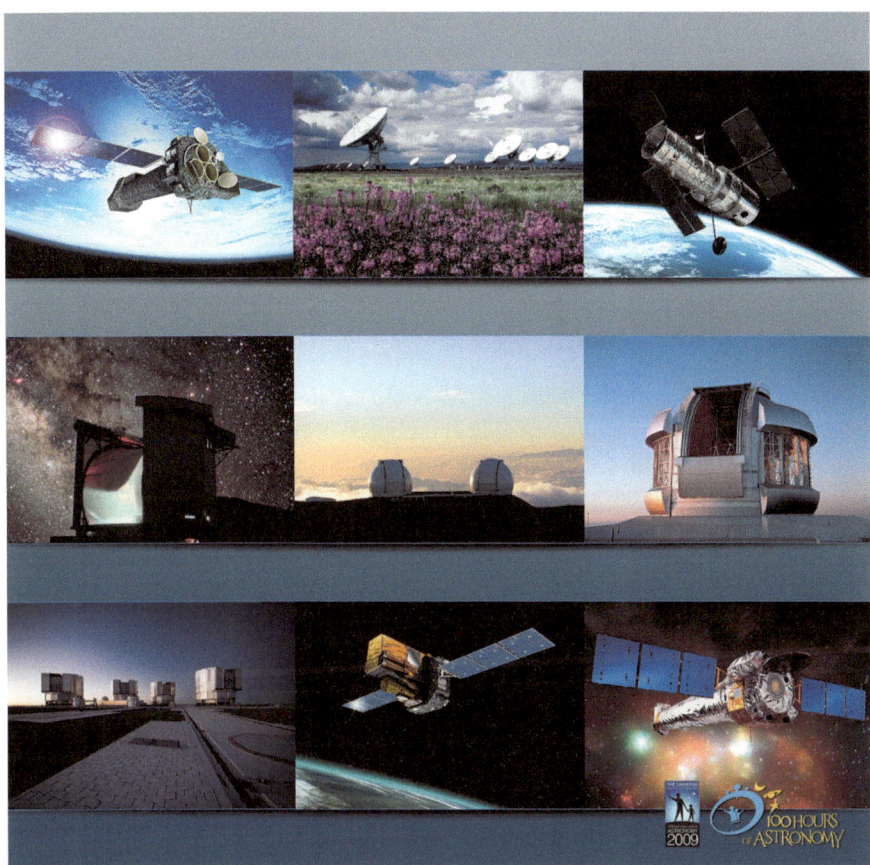

Fig. 6.27 From *left* to *right*, then *top* to *bottom*: XMM-Newton, Very Large Array (VLA), Hubble Space Telescope; James Clerk Maxwell Telescope, W. M. Keck Observatory, Gemini; Very Large Telescope (VLT), INTEGRAL, Chandra X-ray Observatory. (Image courtesy of XMM-Newton/ESA; NRAO/AUI (Kelly Gatlin, Patricia Smiley); ESA/Hubble; Nik Szymanek; W. M. Keck Observatory, Polar Fine Arts/Gemini Observatory/National Science Foundation (Neelon Crawford), ESO, INTEGRAL/ESA/D. Ducros; Chandra X-ray Observatory/NASA)

dreams of seeing the cosmos crystallize into reality. Its discoveries provide a sense of constant wonder about the origin and the future of our universe, its complexities and mysteries. The Hubble Telescope is everyone's telescope; it is the telescope of the people. It has kindled and fired our deepest passions about astronomy and has given us hope that there is more to life than the mundane pathways that sometimes characterize modern society. The Hubble has proved that whatever we can imagine, we can build. Whatever we do not understand today will be understood tomorrow. With the universe as our laboratory, there is so much more that we can contemplate, ponder, and accomplish (Fig. 6.27).

However, it is still worth a trip to visit a land-based observatory. Why?

Why Make the Trip?

Other than the obvious, what reasons are there for making a trip to an observatory?

- Combining leisure travel time with learning is a great way to feel fulfilled during your time away from it all. There's never anything wrong with a vacation that includes plenty of beach time, but it becomes all the better if it's possible to add a little fun education in at the same time.
- The opportunity to interact and ask questions of professional astronomers.
- You may be able to combine the trip with some of your own observing under dark skies.
- You have the chance to view history. We could say that the progress of technology and humankind in general is moving at the speed of light. One day the telescopes of our era will become dinosaurs, so to speak. We have the chance to see them now before this era has passed us by.
- You give yourself the opportunity to be around other people who really enjoy astronomy. There is also the prospect of forming new friendships with some of them.

Okay, so you want to make the trip, but the rest of your family may not put visiting an observatory at the top of the list of how to spend vacation time. If you cannot convince everyone, you don't have to go it alone. You may be able to find a travel partner from your astronomy club, or maybe a group from the club could travel with you. Traveling as a group is fun and may open up the way for certain discounts not enjoyed by just one or two persons. Also, clubs are often looking for new and interesting material for meetings. Sharing highlights of a trip to an observatory could make for a good change of pace at the next meeting. You could include some pictures and video as well. Some of your travel highlights may also be great for use in an outreach program. For example, information could be shared from your own recent experience of visiting an observatory in a location that is free of light pollution. This is a great way to impress on others that we all can and should help preserve those great dark skies.

Traveling Tips

No matter what your travel destination may be, a warm beach, snow-capped mountains, or a city thriving with night life, there are certain preparations and concerns that should be considered ahead of time. Planning a trip to an observatory is not any different. You will definitely need to plan ahead to make the most of your time and to be able to gain the most enjoyment. There is no way that one reference source can cover everything you need to consider. Everyone is different, so there will be specific preferences and needs that one person has that is not so important to another. However, what is presented here should give you some practical ideas and get your own thoughts flowing as to what makes sense for your personal circumstances.

One of the first things to consider is how you will travel. Will you fly or drive a vehicle? There are pros and cons to each decision. Flying to your destination is certainly faster, but you may need a rental car to get around in. Observatories are often located high in the mountains, where public transportation is not accessible. At times shuttles are offered, but that is usually after you have already reached a point fairly close to the observatory itself. Also, you will be more limited on items that you may wish to take with you. Although some 'scopes are compact enough to take on an airplane, for one reason or another this is probably not going to be an option for most people. Trains are another option for travelers who do not have any time constraints. Although much slower they do offer a great way to see the countryside. Traveling by train can also be very relaxing, and the long trip will give you plenty of time to catch up on your favorite books, whether astronomy related or otherwise.

Traveling by car can be cost effective and convenient. As observatories are often located in very out of the way areas, be sure your vehicle is in good working order (and bring emergency items for your vehicle), or you could simply rent a vehicle. If you want to combine the tour of an observatory with your own observing, using a car has the added benefit of providing more room so that you can pack your telescope. If that is one of your main reasons in choosing to travel by car, aside from that essential item, be sure to pack as light as you can. A lot of extra luggage can quickly become unwanted luggage, which can easily put a damper on your trip, or at least become a nagging inconvenience. Some observatories are located many miles away from restaurants, so this is something that you should inquire about before you travel. Does the observatory serve food or snacks? Where is the nearest restaurant? If suitable eateries are not close by you may want to pack some simple non-perishable foods that you can take along.

You should also inquire about the cell phone service. This can be done by checking with the observatory or your cell phone provider. If cell phone service is sporadic at best, find out if the place where you're lodging has Internet service or WiFi. Some people have been surprised to find that they cannot communicate via cell phone at their destination. However, even the most remote places have Internet, so at least you will be able to use email or social media.

This point leads us to the question of where will you stay? In some cases you may not have the typical choice of checking into your favorite luxury hotel, so be willing to stay in a cabin, bed and breakfast, guesthouse, or at a campground. Another consideration is disabled access. There is often at least one kind of tour or visitor program that requires a lot of walking or the use of stairs. Be sure to inquire ahead of time as to the accessibility of everything you wish to see. Check with the facility manager about use of a wheelchair for any disabled person in your group. Accommodations can usually be made as long as you request that assistance in advance.

How will you pay for your trip? Some like to just bite the bullet, so to speak, and swipe a credit card. Others prefer to save. If your trip is some time off, try saving a little each week by cutting out little non-essentials, whatever they may be. For some that just means saving a few dollars by removing soft drinks or expensive beverages from their weekly expenditures. Have you recently upgraded your telescope? If so, you may be able to sell some of your old astronomy equipment. One additional important consideration. Find out what other things there are to do. Go and see some of the other sights nearby. There may be opportunities for visiting amusement parks or museums, and activities such as skiing, golf, tennis, horseback riding, hiking, and biking, to name a few. Especially is this the case if other members in your family do not have astronomy as their primary interest in going on the trip. If this is the case, then you will probably want to fly in so as to be able to have as much time as possible for other activities.

As an example of what you can expect at many observatories, let's briefly take a look at one located in the United States, the McDonald Observatory.

The McDonald Observatory is well known for its astronomical research in connection with the University of Texas. It produces the StarDate radio program, which has been making the public aware of astronomy daily since 1978. It is also heavily involved in community outreach programs and in efforts to limit light pollution. The McDonald Observatory is located just outside of the small town of Fort Davis, Texas, and is 145 miles north of the Big Bend National Park. Big Bend is one the International Dark-Sky Association's recognized gold level dark-sky parks and has a Bortle Scale rating of 1, which essentially means that just about nothing is going to hinder your view of the night sky. The limiting magnitude for stars is 6.8, and the Milky Way and faint meteors are routinely visible in the dark skies there. Although the McDonald Observatory is a little north of Big Bend in the Davis Mountains, the rating of the skies there is still very high at 2 on the Bortle Scale. The closest commercial airport to the observatory is in El Paso, about 185 miles away. There is no cell phone service near the observatory. An on-site cafe offers a variety of food selections for less than $10. Reservations can be made in advance for lodging that is available on-site.

The 82″ (2.1-m) telescope at this observatory was completed in 1938. At the time it ranked as the second largest telescope in the world. In 1966 it was named after Otto Struve, the first director of the observatory, from 1932 to 1947. It has received remodeling and updating over the years and is now computer-controlled. It is also one of the largest telescopes open to public viewing (Fig. 6.28).

The 107″ (2.7-m) Harlan J. Smith Telescope was built in 1968. It was used in planetary study and also was used in lunar laser ranging to establish an accurate distance to the Moon. Harlan J. Smith was the observatory's director from 1963 to 1989, and launched the astronomy program at the University of Texas. The Hobby-Eberly Telescope is a large 433-in. (11-m) instrument for spectroscopic study and was completed in 1996. Its honeycomb-shaped primary mirror is made up of 91 hexagonal mirror segments and weighs 13 tons by itself (Figs. 6.29 and 6.30).

During the daytime the staff offers a solar viewing program that last about an hour. Additionally there are other astronomical exhibits available in the Visitors

Fig. 6.28 The 82-in. Otto Struve Telescope. (Image by the author)

Center. Another daytime tour is provided for viewing the 107″ Harlan J. Smith Telescope and the Hobby-Eberly Telescope. That tour lasts about an hour and a half or more. Children five years of age and under are free, other tickets for the tours are $12 and below. Two to three times each week a strict lights out (to maintain dark adaptation) star party is offered, which usually lasts for about two hours. It begins with an educational Twilight Program that takes audiences on a tour through the constellations. Then guests can move on to viewing the night sky through various telescopes. On cloudy nights indoor presentations are offered featuring virtual sky tours, videos, and demonstrations (Fig. 6.31).

The McDonald Observatory also offers special viewing nights. During the night tour of the 107″ telescope, the focus is on understanding astronomical research (such as spectroscopy) and includes discussion with a professional researcher. Guests also get to observe up to three objects through this telescope. This tour lasts four hours and dinner is provided. Tours of the 82″ Otto Struve Telescope is one of the most popular. Guests are able to view several more objects through this telescope during the course of a 3.5-h tour. A third tour is available of the 36″ telescope and is also a 3.5-h program. This telescope is more readily available during the year and the cost is about half of the other two programs.

There are so many other observatories that can be mentioned worldwide such as: Mount Wilson Observatory in California, Kitt Peak in Arizona, ALMA in Chile, the

Fig. 6.29 The 107-in. Harland J. Smith Telescope. (Image by the author)

Royal Observatory in Greenwich, Mauna Kea, in Hawaii, and many more. The StarDate website has organized a list of many observatories that offer public viewings and other programs. The web address is: http://stardate.org/nightsky/public.

Fig. 6.30 The 433-in. Hobby-Eberly Telescope. (Image by the author)

At the present time we cannot visit any observatories in space, although in the future that may change. Already, innovators of commercial space flight are working on people being able to have access to space travel to places such as Mars. It is conceivable that one day this may become common place. Until then, however, an educational and fun trip to one of many observatories located around the world can be planned. Learning from these great centers of astronomical science can help you to continue making a success of astronomical observing.

Summary

There are several ways in which amateur astronomers can help to promote the preservation of our dark skies. Educating the public at outreach events is a key way to do so. Another tool to help is simply participating in the Astronomical League's Dark Sky Advocate program.

Since 1988 The International Dark-Sky Association has been leading the way in promoting better ordinances and lighting practices to help limit light pollution. This problem affects far more than our ability to view the stars. Light pollution adversely

Fig. 6.31 The Frank N. Bash Visitors Center at the McDonald Observatory The sundial court is visible in front of the building. (Image by the author)

impacts the health and well-being of humans and animals. It also causes a tremendous cost in the form of energy that is wasted because of either poorly constructed lighting or lighting that is not aimed at its intended destination.

Observing at a truly dark sky site is one of the best ways to learn the sky and to be able to enjoy everything that visual astronomy has to offer. There are many such sites still available throughout the United States, Canada, the United Kingdom, Europe and other places due to the vigorous efforts of people concerned about protecting the heritage of our pristine dark skies.

Visual astronomy is an art that has been enjoyed by humankind for thousands of years, from the time when we could use nothing more than the eye up until the time of Galileo when he introduced the telescope to astronomy. Since that time over 400 years ago, our ability to understand the universe we live in has increased exponentially. With the launch of so many space-based observing instruments, there seems to be no limit to how far our understanding can reach. Telescopes such as Hubble, Spitzer, Chandra, Herschel, and others have provided the most stunning and revealing look at what exists as far out as the current edge of the observable universe. However, capturing images of the cosmos is not limited to space-based observatories. After enjoying all that a clear night sky has to offer, you may be wondering how you can preserve its memory. One way is through photography, which is the subject of our next chapter.

Chapter 7

Capturing the Cosmos: Getting Started in Astrophotography

It was one of the most important breakthroughs in the history of astronomy, and it occurred less than 200 years ago. Since the fifth century B.C. people have experimented with the process of capturing light. But it was in 1826 that French inventor Joseph Niépce (1765–1833) became the first person to capture a permanent image from a window in France. It was an 8-h exposure, and although rough, it set innovators on a path that would lead to a significant change in our understanding of the universe. Niépce soon formed a partnership with Louis-Jacques-Mandé Daguerre, and working together they continued to experiment with various methods of creating permanent photographs. Unfortunately Niépce died in 1833, but Daguerre continued his work and in 1839 discovered a way to combine silver iodide and mercury vapor to provide sharp images.

Astronomers soon realized the high potential that photography would give to their study of the stars. On July 17, 1850, the first image of a star was taken by J.A. Whipple and William C. Bond. That image of Vega taken at the Harvard College Observatory would be the first of many, in a race to catalog and study the celestial sky. Many others contributed to the success of astrophotography, such as Henry Draper, Julius Scheiner, William Huggins, and Hermann Vogel.

By 1898 Lick Observatory astronomer James Keeler (1857–1900) began to photographically survey 120,000 galaxies. With surveys like this, over time the evidence began to pile up that would eventually lead to the correct answer to the question of the true nature of spiral nebulae (galaxies). The answer came in 1924 after Edwin Hubble carefully examined photographic plates archived from as far back as 1909. His study of photographs led him to the conclusion that the Milky Way Galaxy is but one of billions in the universe. Astronomers finally grasped that spiral nebulae that had been observed since the time of Messier were external to our own galaxy.

© Springer International Publishing Switzerland 2015
D. A. Jenkins, *First Light and Beyond*, The Patrick Moore Practical Astronomy Series, DOI 10.1007/978-3-319-18851-5_7

Photography has come a long way since then, and there is good reason to believe that it will continue to become better in terms of resolution, speed, and effective storage and transmission of photographic data. People love astrophotography for various reasons. Some simply enjoy the highly detailed views that cannot be gleaned with the eye alone. There is also the aspect of art and astrophotography. Some use images of celestial objects in mixed media art. On the other hand there are also many uses for astrophotography beyond simply enjoying the aesthetic beauty of deep space color imagery. Some amateurs use it to do serious research that in the past had normally been done only by professional astronomers. You may have heard about many amateur astronomers that have been noted as a sole or joint discoverer of a supernova. In many cases this is a result of imaging multiple galaxies several times each month. After carefully comparing new images to older ones, a discrepancy in the images are sometimes identified as a new supernova. Other ways imaging can be used for research involves photometry and spectroscopy. In 2012, a high school student from Dripping Springs, Texas, used a homemade spectroscope, CCD (charged-coupled device) camera, and other equipment to analyze two eclipsing binary star systems. For her topnotch research, she won first prize in the Priscilla and Bart Bok Award presented by the Astronomical Society of the Pacific and the American Astronomical Society.

Still other amateurs are using imaging to contribute valuable information that helps to discover exoplanets. The equipment they use is a telescope with a clock drive, computer software, and a CCD camera. There are not enough professional astronomers to observe and monitor changes in stellar brightness that may indicate the presence of a planet. Also, several times each year NASA promotes exciting activities involving imaging in partnership with the Astronomical League. Participants must image an object such as a comet or a planet during a specific time period. Then the participant must engage in some sort of outreach event connected with the object. Some examples of these periodic observing challenges are Comet Siding Spring's close approach to Mars, Comet C-G, and Pluto. More information about these imaging challenges can be found by clicking on "Observing Challenges" at: http://solarsystem.nasa.gov/news/. However, does this mean that you must have a CCD camera in order to do any astro-imaging at all? Getting started does not need to be either expensive or complicated. You need a few pieces of basic equipment, patience, and ingenuity.

Digital Cameras

Point-and-Shoot

Most everyone owns a point-and-shoot digital camera. You may wonder if it is suitable for astrophotography. Generally speaking it is not, because although many have a high number of megapixels, the size of the pixels are too small for the high resolution needed for astrophotography. Most have pixels that are 2 μm in size, whereas DSLR (digital single lens reflex) cameras have pixels that are 4–8 μm in size. The result is an image with a lot less "snow" or noise in it. However, this does not completely rule out using a point-and-shoot camera. You have to start somewhere, and it will give you a good chance to experiment.

Actually, some people have taken shots of the Moon and planets with a handheld image stabilizer point-and-shoot camera with fair results. A better option would be to use a tripod to steady the image, but if you do not have one try resting the camera on a flat surface and using the timer so that the vibration from pressing the shutter does not ruin the image. Set the camera to its lowest f ratio, and adjust the ISO setting to a higher number (but not too high since it will increase noise in the picture, ISO 400 or ISO 800 would work). The best way to see how it will perform is to try out multiple shots. Take several shots, varying the exposure time from 1 second (s) up to no more than 30 s.

Another way to stabilize a point-and-shoot camera is to use an adapter to attach it to your telescope. These can still be found through companies that sell camera equipment. One such company is Telescope Adapters (www.telescopeadapters.com). They sell kits that come complete with everything you need to attach a point-and-shoot camera to your telescope so that you can shoot images through the eyepiece. Using the listing on the website, look up the manufacturer and model of your camera, and the site will indicate which kit you will need. This method of photography would be only for objects that do not move very fast, in a low magnification eyepiece, so that pretty much limits you to shots of the Moon. After taking a series of shots use an image-processing software to improve the quality by decreasing the image size, making any imperfections less obvious. You can also adjust the color in the image, contrast, and brightness, along with a host of other tweaking.

DSLR Cameras

If you really want great images, a DSLR camera is required. With your DSLR camera mounted on a tripod, you will be able to take great shots of the Milky Way, constellations, planets, and the Moon (Fig. 7.1). Just as in the case of a point-and-shoot camera there are different ways of gathering images, but in every case a DSLR has a superior performance. Most astrophotographers will agree that Canon provides the best cameras suitable for astro-imaging. When using your DSLR on a tripod, you will need to keep your shots at 30 s or less. The reason is that Earth rotates, and so the stars will appear streaked in the image. If you want a photo containing star trails then this is a good thing. Many astrophotographers have done some interesting work with star trails. Adjust your camera to a higher f ratio and take a longer exposure, perhaps for 5 min or a little longer. The longer the exposure time, the longer the star trails will appear. No matter what kind of shot you take, a cable release is an accessory that is essential to limit vibrations. A company called Vello offers wireless remote control for specific Canon DSLR cameras.

Fig. 7.1 Quality photos of the Milky Way can be captured by simply using a camera mounted on a tripod. (Image courtesy of ESO/C. Malin [christophmalin.com])

DSLR cameras are sensitive to red light but normally have a filter inside that blocks infrared light so that photographs taken in the daytime look normal. However some digital cameras can be purchased that have been modified to have twice the sensitivity to the hydrogen-alpha line. This is the wavelength common in nebulae such as the Orion Nebula, the Horsehead Nebula, and many others. The difference in the detail captured by this kind of modification is impressive. For some, this feature is more important than a DSLR that also can be used for daytime photography. One company that offers these modified cameras is Hutech Astronomical Products.

There are several settings on the camera that will need attention. Most DSLR cameras have a live focusing feature. This allows you to bring stars into focus right on the display screen. There will also be a choice of what format in which to save your images. In any event, be sure to select RAW images along with JPEG. Having the image saved in RAW format will give you more options for image manipulation later. The JPEG file format is okay, but because of compressing the file much of the original image data is lost.

The inherent noise in digital images can be limited in two ways before processing images in a software program. Before buying one it would be good to check for these options. The first is a noise reduction feature available on some cameras. Another feature is the ability to take a dark frame exposure along with regular exposures. During a shot some pixels are hot, meaning that they have a pattern of noise that is usually the same each time. These unwanted noise patterns are caused by varying temperatures, exposure times, and tiny imperfections and variations in the pixels. By taking dark frames with the sensor covered, these unwanted noisy pixels can be subtracted out of your images. This used to be a separate task that would have to be performed along with regular images. Then these dark frames would be subtracted manually or with a software program that can do this for you. *DeepSkyStacker* is one such program that works well and is available free of charge.

However, some cameras now offer an inherent function long-exposure noise reduction feature. This automatically takes a dark frame after each exposure and then subtracts it from the image. Another effective method of eliminating noise from an image is to take several short exposures and stack them together. Instead of taking one 30-min exposure, 30 separate 60-s exposures can be taken. Then all of the shots are added together. This gives a cumulative effect providing more freedom from the distorting effects of noise. However, the best shots still come from longer exposures.

Taking Shots Through the Eyepiece

Just as in the case of a point-and-shoot camera mentioned earlier, a DSLR camera will also need an adapter to be able to take shots through your telescope. Also, once you attach everything your 'scope will likely need to be rebalanced for shots of the Moon and very short exposures of the planets (1/100 s). For longer exposures and to image star fields through the eyepiece you will need a telescope that tracks.

We often see advertisements for tracking mounts that are expensive, but there are a lot of good and affordable options. The major manufacturers such as Celestron, Meade, and Orion offer affordable mounts that are great for small apochromatic refractors (80–110 mm). These are particularly good for imaging, as they are very portable, need shorter exposure times for images, and offer a wide field of view (for imaging nebulae and large star fields).

For focusing during imaging of deep sky objects, the live view on DSLRs works best. However, some prefer to hook up their camera to a laptop or video screen, where they can use the larger view to properly focus the object. However, this is more equipment and cables to keep track of, and increases the need for a portable power source. After imaging for long periods of time, many astronomers have had to call it a night because their batteries went dead. Be sure you have extra batteries or another sufficient power source.

Another way of getting great shots is through a webcam. These are now sold by companies complete with the necessary adapters to work with your telescope. Webcams provide very clear shots of the Moon and planets, and feed data directly into your computer. With video you can capture every moment of the night, which includes moments of especially good seeing or some moments that are not so good. The great thing about it is that you can collect only the good frames that were shot and have the computer software edit them together into one movie or stack them together into an image.

Using Piggyback Mounts

The DSLR camera is sensitive enough to take great shots without being directly attached to the telescope's focuser or eyepiece. The problem is tracking images during long exposures. Many therefore prefer to simply piggyback the camera on top of the telescope that is equipped to track the motion of the stars. A mount that tracks in altazimuth mode will have stars that appear to rotate around the center unless you take very short exposures. Therefore to successfully image using a piggyback mount you must use an equatorial mount that has been precisely polar aligned. The most convenient way of ensuring this is use a mount that has a polar alignment 'scope in the mount. If your mount is stable and has been accurately polar aligned, exposures of up to 10 min or so can be taken without guiding.

In longer exposures guiding is necessary because of periodic errors in a mount's drive gears and occasional star drift. Also, if the 'scope is not precisely polar aligned, this will cause more drift. Air turbulence can also cause small movement in the position of an image. This is true whether you are shooting with a piggyback setup or through the focuser.

Guiding can be done manually by using a guidescope equipped with an illuminated reticle eyepiece with crosshairs. The corrective motions are done with fine motor motion controls on both axes. This is more affordable than using an auto-guider, but more tedious because of the constant sighting and manual adjustments that must be made during several minutes of a long exposure. However, if you do choose to use manual guiding, after watching carefully to see in which direction the guide star is drifting, rotate the eyepiece so that one set of the crosshairs are lined up parallel to the star's motion.

One type of auto-guider connects the guidescope to a computer. The software is triggered when the guide star moves off center and automatically adjusts the position of the mount to stay on track. Another kind of auto-guider connects directly to a mount that is already designed to communicate with an auto-guider. Many manufacturers now offer these. Celestron's NexGuide auto-guider connects to any mount that is equipped with a port to accept an auto-guider and also connects to the guidescope. It uses a CCD sensor capable of locking onto a star as faint as magnitude 8.0. The device also has a night vision view screen to keep track of the status of the auto-guider. At under $300, this and others like it offer a way to have this feature at a reasonable cost.

Using a Camera Tracking Mount

If you do not wish to piggyback your camera on your telescope, there is another great option available so you can make images that track the motions of the sky. Many amateurs are mechanically inclined and have built their own tracking platforms and mounts. But for those of us who are all thumbs in this area, there is a solution right out of the box. One tracking mount that has good reviews is the iOptron SkyTracker. The device attaches to most camera tripods and has a polar alignment 'scope for accurate tracking. To help with alignment they have made an app compatible with iPhone or iPad that makes the alignment process even easier. SkyTracker can hold weight of up to 7 lb, more than enough for most cameras. Power is supplied by AA batteries with a 12 h life in cold weather, and more if used in more favorable weather. It can be used in either the northern or southern hemisphere. Simply turn on the 'scope, align it, and start imaging.

High End Astrophotography

For some amateur astronomers, astrophotography is their passion and chosen area in which they specialize. They can be found outside all night if need be just to capture the perfect series of images. That means no shortcuts on equipment—for them only the best will do. To acquire the best images, you will need to take long exposures. Taking long exposures requires guiding (preferably with an auto-guider). Shots that are of a higher quality than those made with a DSLR camera involve using a larger telescope aperture, a very sophisticated mount, and a CCD camera. Some of the companies that offer the high end equipment are SBIG (Santa Barbara Instrument Group), Finger Lakes Instrumentation, TeleVue, Software Bisque, PlaneWave Instruments and others.

Although a GOTO 'scope is not required for astrophotography, it certainly does make sense for several reasons. Many GOTO 'scopes come pre-packaged with many of the features you need for astrophotography, eliminating having to make so many separate acquisitions. A GOTO 'scope is also practical when you want to shoot in light-polluted skies. After a manual or auto bright star alignment your 'scope will be able to automatically slew to whatever imaging targets you desire. Computer-controlled telescopes can also be programmed to image a designated series of targets whether you are sitting down at the computer or not. This greatly enhances quality of life, since you can make other use of your time and still capture the shots that you want.

Image Processing Software

Once you have taken the shots you want with the right equipment and exposure time, there are still a few more steps in your journey towards a beautiful image. Now it will be necessary to process them with photo enhancement software. There are a host of different software tools, and most have great features. Be open to try-ing different ones until you find one that best suits your need. Many astrophotogra-phers prefer *Photoshop*™, as it is relatively easy to use and has lots of options for photo manipulation. However, several are free of charge, such as *GIMP* and *DeepSkyStacker*. Some programs only perform initial image stacking, refinement, and conversion of raw image data into TIFF files, while others are all inclusive programs that can do it all.

ImagesPlus 6.0

One such comprehensive software package is called *ImagesPlus 6.0* offered by Mike Unsold (www.mlunsold.com). This program was favorably received 10 years ago when it was featured in *Sky & Telescope's* list of hot products for 2004. It processes DSLR images, one-shot color images, and CCD images, including functions such as alignment, stacking, and calibration. Additionally for fine-tuning astro-images the software can provide color adjustments, sharpening, smoothing, stacking, and much more. His website offers a wealth of information to support the product, including dozens of tutorials. One of the main advantages of this software is that it is designed specifically for astrophotography, so you'll find that many of its features are intuitive and just make sense in this context.

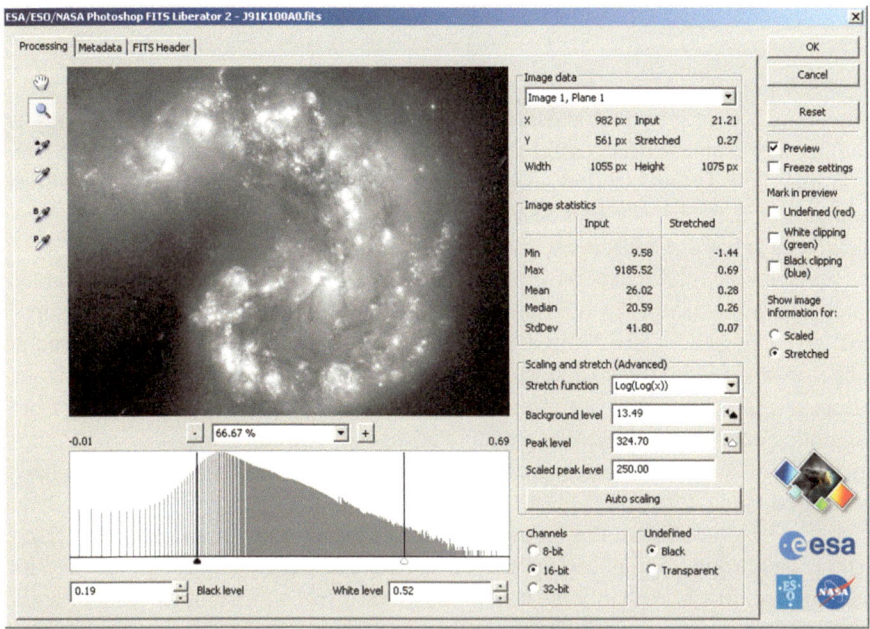

Fig. 7.2 Screenshot of FITS Liberator image processing software. (Image courtesy of NASA/ESA)

FITS Liberator

FITS stands for Flexible Image Transport System. FITS is a raw format used for astronomical image data. Files that are stored in the FITS format can be easily manipulated in the *FITS Liberator* and prepared in that system for further refinement in a program such as *Photoshop*. It was created in 2004 to facilitate the creation of images from the raw data received from various telescopes such as Hubble, Spitzer, and others. The latest version is available for download from the Space Telescope website free of charge. The program allows entry of metadata for each image, offers advanced options for image enhancement, including noise reduction, and provides a fast processing of images due to better memory management. Data can be saved in either an 8-bit, 16-bit, or 32-bit channel. After enhancing an image, the file can then be saved in the TIFF file format for further enhancement in the program of your choice. A common function would be to use *Photoshop* to create a composite color image. The application is designed to run on both Windows and MAC (Fig. 7.2).

Astrophotography Workshops

Attending a workshop that is especially designed for learning astrophotography is another way to ensure a successful start. Check the goings-on in the city where you live. You may be surprised to find such a class. Also, there are many such workshops being offered at dark sky sites that offer the most beautiful views of the Milky Way as well as stunning landscapes.

For some people, hands on instruction offered in a workshop under the stars is the best way to learn. An Internet search for "best astrophotography workshops" will give you quite a few choices. Astrophotography workshops are also offered at some observatories. For example, the Kitt Peak National Observatory in Arizona offers several 3-day astrophotography workshops throughout the year, ranging from beginning concepts to those that are more advanced. You can bring your own equipment or use equipment that is provided, and you get to take images that you have processed home with you.

Let us also not forget that in every astronomy club there is at least one person who is an astrophotographer. If you need someone to walk you through the process step by step, this is probably the best place to start. Just as with other parts of amateur astronomy, it can save you a lot of time, and you can observe how others who are successful at it use their equipment to achieve the best results.

Internet-Controlled Astrophotography

For those who do not wish to spend the additional money needed for astrophotography there is another option. You can rent telescope time and control a telescope remotely. This is not a new concept; there have been some companies offering this service for several years now. However, with increasing Internet accessibility and higher Internet speeds, this has recently become appealing to more people. No need to learn how to set up costly equipment, no need to repair or otherwise maintain it. Everything is handled by the hosting company. You only need to know how to use a computer and learn the hosting company's software interface that controls the telescopes.

This service gives amateurs the best of both worlds—you can have the satisfaction of collecting your own image data, process it yourself, but not have to pay full price for it. You pay only for the time spent shooting the images you want. Another advantage is that you can sit in the comfort of your own home and never have to set one foot outside to do imaging. This is an important advantage during cold winter nights, when traditional astrophotography often requires spending at least some time outside (unless you have a sheltered observatory).

At this time, one of the best companies offering this service is iTelescope (itelescope.net). There are several reasons why this is the case. First, iTelescope has excellent customer service. In this Internet age, all too often we have difficulty being able to communicate directly with people anymore. However, the people at iTelescope are very good about communication. They will respond promptly to your emailed requests. Secondly, the iTelescope website is very well designed and easy to understand. There is a wealth of information already there that explains exactly how to use their equipment. Typical questions are on the FAQ page or on the tutorial pages. Several video tutorials are also available on the site to provide guidance on using their telescopes. Another feature iTelescope offers is a monthly newsletter. It provide updates, tips, coupons, and articles on imaging.

Payment for the service is provided by a point system. Each of the 19 telescopes is assigned an hourly rate (in points). Customers subscribe to the service for whatever amount is desired on a monthly basis. The higher monthly fee you pay for a subscription, the lower the hourly rental fee is in points. The good thing about this system is that iTelescope only charges points from your account for actual imaging time. You are not charged for simply being logged into the site, or even for the time when the 'scope is slewing to an object. The rate per hour for each telescope is also governed by whether or not the Moon is out. Depending on the amount of illumination in the sky from the Moon, a discount is offered that varies from 10 % to 50 %. Of course, moonlight can adversely affect imaging, so you must keep this in mind when choosing to image during those times.

To help customers plan a session of imaging, iTelescope provides weather updates through their website. This allows hour by hour tracking of weather and related observing conditions such as seeing and transparency. Live cameras feed into the website, providing views of the skies where the telescopes are actually located. Another great planning tool is their reservation system. It would be very disappointing if you planned to image a comet on the last night possible only to find that all of the telescopes were already in use by other imagers. That's where the reservation system comes in.

Making a reservation is done through a simple and intuitive calendar-based interface. You first select the telescope you want to use, then you select the hours on the day you want to reserve. A reminder is sent via email to alert you about upcoming reservations. Once your scheduled time arrives, simply start selecting targets to image for the duration of the reservation. Another excellent feature in connection with the reservation system is the ability to program a series of images you wish to take ahead of time. So once you tell the system to execute a plan, you do not even have to be at your computer; the system will automatically image the requested targets at the specified times.

On the rare occasion that there is a system malfunction iTelescope does not charge you for the time because no images were made. If for some reason an image was taken but was ruined because of some malfunction a refund is offered. This also is the case if there is a sudden closure of the observatory due to inclement weather; no charge is made to your account since the observatory had to close. Reservations do not cost any additional funds, it is all a part of each customer's subscription package.

New customers can sign up on the iTelescope within minutes. The site is clearly marked with instructions on how to sign up with its inviting "try it for free" icon. True to their word they do offer a no obligation free trial. A basic amount of points are deposited into your account, and you are allowed to use some of the more basic telescopes. Once you decide to become a regular member, rates start at approximately $15 USD per month and up. Customers always have the option of upgrading their account or downgrading it. The higher the monthly fee, the lower the rental cost in points for each telescope.

iTelescope allows members to image any target desired. The object does not have to come from a known database. As long as you know the coordinates where you want the telescope to be centered, you can image. This opens up a lot of possibilities, including supernovae, asteroids, comets, and other transient astronomical objects or events. The company also allows use of multiple telescopes at the same time, so you can have several projects going on at once if the systems are not already being used by others.

If you live in the northern hemisphere and have always wanted to image southern hemisphere targets, you can now do so through iTelescope. They have 19 telescopes available in three different locations: New Mexico in the United States, Nerpio, Spain, and New South Wales, Australia. The apertures range from 106 mm to 510 mm. The equipment offered is very sophisticated, offering all kinds of options including luminance, color, narrowband filters and much more. Images are captured in a JPEG file as well as in the raw FITS file format so that images can be manipulated manually to get the most out of the captured data. All images taken with their telescopes are stored in your account for 30 days before removal. Simply download your file data to a hard drive or a cloud drive to be sure nothing is lost.

One of the best reasons to use iTelescope is that all images taken by the customer belong to the customer. If you take an image, you own the image. No extra fees, no requirements to post your images anywhere, just straightforward imaging and retention of ownership. All things considered, iTelescope provides an alternative form of astrophotography all from the comfort of your own home.

For all that astrophotography has to offer, what we've considered is just the tip of the iceberg. After you start trying out some basic techniques, seek out the astro-imaging community online to keep up to date on the latest methods and equipment. Astronomy magazines also cover new methods of imaging at least on an annual basis. Just as in the case of visual observing, be patient with your burgeoning astrophotography skills; it takes most people years of practice to perfect the art of imaging. It's not just about the right equipment. It's the correct use of those tools along with smart image processing afterwards. Try imaging for different reasons so as to keep your mind fresh with ideas and open to using new techniques to achieve great results. Some of the ones mentioned here were art, observing faint details in objects, research in spectroscopy, or searching for supernovae and exoplanets.

We learn by doing, and with a healthy sense of humor you will be able to take any poor images you shoot with a grain of salt and learn from missteps. Ask questions of those around you (online or in your astronomy club) that are more experienced. Make the learning curve fun by entering astro-photo contests and submitting your work to magazines or online. Simple activities like this will help you to stay motivated and will generate improvement. Eventually you'll be able to shoot the stars like the pros, have a collection of images you can share and be proud of, and continue to find new ways of enjoying amateur astronomy.

Chapter 8

Finding the Best Astronomy Reading

Within the pages of history, Galileo Galilei will long be remembered among those scientists who relied on observations rather than the assumption and conjecture common in his day. Galileo's passion for discovering the truth about the world around him moved him to point his telescope towards the sky in late 1609. With his telescope working at 20×, he began to make detailed observations of the Moon and the stars of the Milky Way. Along with his observations of Jupiter and its moons, these efforts marked in our history a new age in which humankind would finally leave behind the view that Earth was the center of the universe. The mathematics professor from the University of Padua would collect together his observations along with detailed illustrations in a book (*The Starry Messenger*), published on March 13, 1610. In just one week all 550 copies of this historic work were sold, and thereafter it became one of the most noted astronomical publications ever produced.

Who exactly first invented the telescope is still somewhat in dispute. Most historians accept Dutch lens crafter Hans Lippershey in 1608, or his competitor Zacharias Jansen perhaps as early as 1590, although this is doubtful. In any event, Galileo was probably not the first person to point a telescope towards celestial phenomena. For example, Thomas Harriot is known to have observed the Moon and perhaps other celestial objects with a telescope as early as the summer of 1609. It is clear, however, that Galileo was the first to record and publish his findings based on his visual observations using this relatively new instrument. Here at last, in the pages of *The Starry Messenger*, was a written work, complete with drawings, of his telescopic observations including the Moon, Jupiter, and the stars of the Milky Way. With it he confirmed ideas that had long been promoted by Nicholas Copernicus. It was a work that would be referred to again and again, moving still others to make their own observations and take astronomy further than anyone in the seventeenth century had ever imagined.

© Springer International Publishing Switzerland 2015

D. A. Jenkins, *First Light and Beyond*, The Patrick Moore Practical Astronomy Series, DOI 10.1007/978-3-319-18851-5_8

The written astronomical findings of Galileo and contemporaries was just the beginning. Soon publications began to accumulate and abound, written by great minds such as Johannes Kepler, Edmond Halley, Isaac Newton, and many more. Fortunately for us, these great works have been collected and archived.

In 1820 the Royal Astronomical Society was founded by a group of men including the famous astronomer William Herschel, and it was chartered in 1831. Today its library has more than 35,000 bound items, including books and journals, along with many more unbound pamphlets. It is one of the largest collections of rare astronomical books in all of Great Britain. The RAS library also includes a vast collection of photographs, portraits, illustrations, instruments, and relics. One characteristic that demonstrates the value of this collection is that it contains many rare gems in the history of printed astronomical knowledge. There are myriads of great rare books, but here are a few that stand out, especially in the context of our discussion of successful observing.

The RAS library has a first edition copy of *De Revolutionibus*, the famous book published by Copernicus in 1543, a landmark in the beginning of our understanding of a heliocentric Solar System. The library is also privileged to hold a copy of Johann Bayer's *Uranometria* from 1603, the first to use a grid system and to label each star with a Greek letter, still in use now and known as a star's Bayer designation.

Another fascinating part of the archives of the RAS are the observation notes of William Herschel. One such notebook dates back to 1781, wherein his original notes can be seen that were made during his discovery of the planet Uranus.

Perhaps the most pivotal book in the RAS collection is a 1653 copy of Galileo's great work *The Starry Messenger*, originally printed in 1610. We are very fortunate to live in a time when we still have access to such a rich written astronomical history!

In our twenty-first century we have at our disposal many excellent sources of information in either digital format or on the printed page. Printed books are still a widely used, practical tool for study and quick reference that is not subject to the need for electrical power or a connection to the Internet. That is not to say that one method of access is preferred over another—both have their own advantages. Furthermore, more and more people are now in a position to almost always have access to the Internet or to some sort of electronic storage device, be it a computer, tablet, e-reader, or smart phone. A book is still an excellent way to learn, and its physical presence is a reminder to continue your exploration of astronomy. Since humans are prone to forget things, books provide important reminders of knowledge, directions, and recommendations. At times some of the information assimilated from a book may not be needed initially, but later on may prove to be extremely valuable.

Read with a Goal in Mind

A quick search on the Internet will yield 30,000 or more results in astronomy books for sale, but how many of those will really benefit you? When trying to locate good reading material it is wise to first be sure that you have established what you wish to accomplish. Although this may seem obvious, a common behavior is to simply browse reading material, without having a specific goal. Of course there is nothing wrong with spontaneous reading or research. However, failing to narrow things down a bit may lead to a lot of time perusing books and magazines without actually finding what you need. Then, after spending precious hours considering information that may not be relevant, you may feel like giving up because of feeling over-whelmed. Therefore, set some goals for yourself so that you can maximize the time you spend reading.

As an example let's say that you wanted to become more adept at recognizing at least a few stars in each constellation. There are many books that feature a discussion of the constellations, but if you are not specific, you may end up reading more on the history of the legends and myths behind the stars than the stars and phenomena themselves. So what should you look for in this case? A practical book on this topic for our stated purpose would need to have charts within the book for easy reference. The guide should also feature some of the brighter non-stellar objects that can be seen with the naked eye. It also would be nice if at the same time the author did perhaps include some history on the constellations without veering too far from the main goal, which in this case is to help you become familiar with each constellation not just academically but visually. Other examples of goals might be to understand more about the astrophysical aspects of certain phenomena, and to learn all of the deep sky objects in a particular constellation and how to locate each one.

It is natural that you may end up reading publications on astronomy that will have topics that overlap from one book to the next. This is fine, since it is good to review different viewpoints and perspectives that authors may have on the same topic. There will surely be points in one book that are not covered in another. Careful review of the table of contents and each chapter's subheadings will help you to avoid any unnecessary duplication in what you read. If at least two-thirds of the book seems unfamiliar or appears to be written from a different or unique vantage point, then it could still be a good read for you. The best way to see if it will be a book that you will continue to refer to and enjoy is to just pick it up and read a few pages. If its style and information will be beneficial for you it will usually resonate within the first few pages of its content. However, let's review a few additional details that you may want to consider before purchasing that glossy new book.

Narrowing Down Your Choices

The backlist of a publisher can sometimes be a good indicator of finding good astronomy reads. Take a few minutes to examine the list of publications that the publisher has produced, which are now usually available on the publisher's web-site. A publisher with a wide range of books on science and astronomy often has the experience and resources needed to provide some of the best material. There are also excellent publishers who offer astronomy publications and others outside of science. Take a look at the quality of work as a whole. If the works have been well received by others, this is certainly a good place to start. Many publishers of textbooks that are used in institutions of higher learning find many of the best authors for each subject, and then the work will go through a rigorous peer review. A similar process is also followed by publishers of books designed for more casual readers.

Consider who the author is, as well as his or her experience within the scope of the given topic. If the author has written several books, it may be helpful to peruse a few of them to see if you are comfortable with the overall style and format that is used. Often a sample can be found either on the publisher's website or through an online bookseller's website. Take a careful look at the main title, summary, chapter titles and subheadings, introduction, and even the index. For a more thorough preview, find out if it can be checked out from your local library. Many of the astronomy books that are prominently featured online or in brick and mortar stores are those with the most colorful pictures. Of course, beautiful illustrations are attractive, enhance the work, and are even desirable to an extent. However, a book judged solely on the fact that it contains great astrophotography does not necessarily mean that it is a great book for you, or that it will help you meet your particular goals.

A good source for finding goods reads are astronomy magazines, which very often contain one or more book reviews in each of their issues. If you subscribe to one, you will find plenty of helpful reviews in any back issues that you may still own. Otherwise, the websites for the astronomy magazines also post their reviews online.

Your local astronomy club is another a great place to find recommendations. It is likely that someone there has read the book you are interested in and can help you to decide whether or not the book will suit you. Ask others what titles they are reading and if they found them to be practical or not. If your club does not already do this, a book review with audience participation may be a practical feature to add to club meetings. Most clubs already have a schedule for specific topics and a brief (10 min or so) book review of a selection that goes along with the current topic could be very helpful and pertinent to the needs of many members. If your club has a newsletter, it may contain a regular book review.

Here are a few online resources that offer excellent reviews of astronomy books:

1. *Astronomy Now* magazine: www.astronomynow.com/. The database on this site is accessible right from the home page and sorted alphabetically. It contains dozens of excellent reviews.
2. Cloudy Nights: www.cloudynights.com. This site has long been a favorite of amateur astronomers and contains many book reviews as well as forums where books are discussed by members.
3. Amazon: www.amazon.com. By going through the selection tree under books and down to astronomy, thousands of books are listed, and most of them have many detailed comments from people who have read them. This site can provide a good starting point to move you in the right direction.

Further Considerations

Okay, so now you have the book in your hands, but you do not have to read it from cover to cover in order to assess if it's right for you. With just a few minutes (10 min on average) spent reading the book, you should be able to judge its quality and compatibility with your needs and personality.

As you begin reading the first chapter, does the author's voice hold your attention? Each author has a different style, and sometimes there are certain ones that seem to create a better response. If you find that you cannot even get through the first few pages without feeling as if you are going in circles, it may not be a good read for you. However, it is just as important to acknowledge that the author may still have some very important points that are being communicated, and you may have to focus a little bit more on trying to connect with the author's style. Once that connection is made, you may find that, although you were previously never a fan of that particular style, it ends up becoming one of your most treasured books. As the saying goes, "You can't judge a book by its cover." Sometimes you will just have to dive in!

Another important question to consider: *Is the book attractive?* By this we do not mean its cover but rather its overall design, layout, and presentation. If you were going to take a first-time trip to a beautiful country on the ocean, you would probably not be inclined to use a guidebook that did not offer any pictures of your destination, or that appeared as if printed on an old typewriter. Similarly, a book attempting to take you to some of the most beautiful destinations in the universe should have at least some images. The images or illustrations would enhance its message and also offers a way to visualize its descriptions.

This of course does not mean that a book full of attractive photos always indicates that it is very informative. But consider that from a practical standpoint, it is hard for the mind to assimilate detailed information if it's presented in an unattractive way. The book's format should be clear and its information arranged in a manner that is easily readable. In this regard another appropriate saying comes to mind, "Tell me and I'll forget, show me and I may remember, involve me and I'll understand." On the other hand, a book lacking in photos that you might like to see may make up for it in extra content. There are different ways of showing concepts to readers and involving them.

Does the book concisely and clearly present the facts that you are looking for? After all, this is why we read a book in the first place, because it seems to offer a set of facts and details with which we are not entirely familiar, or that would further us in our understanding of what we already know about the topic. We usually seek out a particular book based on the title and summary on the back or inside flap, which should be a good indicator of its goal or intent. However, at times these indicators can be misleading, so be sure to carefully review the table of contents and the index. A careful review of those areas along with the titles of the subheadings throughout the book should reveal the kind of information you can expect to read about. Each author presents the facts in his or her own style, and that goes back to our first point. You will have to decide if the author's voice, or way of presenting the facts, holds your attention and seems clear to you.

Is the book written in language that is understandable? If the astronomical publication you are perusing seems too technical, perhaps full of very complex math or physics concepts that are geared more towards an astrophysics student, then it may not be the right book for you, unless of course you have a solid background in those areas. This is not to be confused with new ideas presented that may at first be somewhat challenging to grasp. The key is whether or not the author breaks down the concepts into digestible pieces and then explains them well. Sometimes a book will discuss certain matters, beginning from an assumed starting point, that are not as difficult to grasp as you might think. It may be that a quick reference to the book's glossary or appendix is all that is needed in order to clear up the issue and allow you to continue reading. In other cases, it may be that you need to start with a more entry-level book before going on to read your initial choice.

Does the book give you a complete story and at the same time leave you with a thirst to learn more? In other words, on some level you should feel a sense of accomplishment and stimulation after reading the work. For the given topic it should lead you from point A to Z in a logical and coherent manner. It does not mean that there isn't anything left to say on the subject, but rather that it has come to a natural conclusion and the book gives the impression of being self-contained. By the time you finish the book, hopefully you are not so exhausted from reading it that you're glad it's over. Instead, a work that is well written will seem like a visit from a good friend, and you are disappointed that it will have to come to an end. Although the only way to be truly sure is to read the book all the way through, by perusing the content thoroughly you can arrive at the correct conclusion.

Key Publications

A review of several choices among star atlases and constellation guides was given in Chap. 3, so we will not cover that here. Instead, let's take a look at five other areas: astronomy magazines, astronomical history, biographies of significant astronomers, astrophysics (both observational and theoretical), and observing guides. This will give you a good overview of excellent material that at this time is still readily available. In turn this should lead you to other great reading selections by the same or similar authors and publishers. Remember to use the suggested guidelines discussed so far to help you determine if any of the selections are right for you.

Astronomy Magazines

Sky & Telescope magazine (www.skyandtelescope.com) is considered one of the best astronomy magazines available today. Its rich history goes back to at least 1929, when one of its predecessor magazines was in print under the title *The Amateur*

Astronomer. Soon this title was merged with *The Sky* magazine, published by the Hayden Planetarium in New York.

In 1940 Charles A. Federer took over the publishing of *The Sky* magazine, merging it in 1941 with another magazine written by professional astronomers called *The Telescope*. The new creation, *Sky & Telescope*, was released in November 1941, being published out of Cambridge, Massachusetts. Federer's goal was to provide an attractive, illustrated magazine that would be indispensable to the amateur and, at the same time, provide a platform where professional astronomers could reveal their research to the public.

Most would agree that for over seven decades this magazine has fulfilled those needs. As of the date of this publication, a print subscription automatically gives a subscriber access to the digital version at no extra cost. With the addition of so much more information on the magazine's website, *Sky & Telescope* continues to be in the forefront in making astronomy accessible to millions around the world.

Sky & Telescope magazine is full of practical astronomy information each month, covering in great detail the deep sky objects and planets that are visible. Its monthly star chart is currently printed on paper of a slightly heavier stock that will hold up to field use. Detailed finder charts are always included to aid in finding comets and fainter planets such as Uranus and Neptune. The monthly segment "Celestial Calendar" by Alan MacRobert always contains fascinating and challenging objects within reach of small- to medium-sized telescopes or at times even the naked eye. If you love to observe deep sky objects you will find yourself turning to the column by Sue French before any others. Well known as an expert observer, her articles in the series "Deep-Sky Wonders" are of the highest quality and include all of the necessary information to successfully locate between 6 and 12 celestial sights each month. *Sky & Telescope* is also well known for covering the hobby of amateur telescope making. The column "Telescope Workshop" by Gary Seronik gives excellent coverage to cutting-edge telescope designs and equipment made by amateur astronomers all over the world. Each issue is also full of great photographs, including many by amateur astrophotographers from all over the world.

Astronomy is another welcome addition to the list of periodicals; it came onto the scene in 1973. Similar to *Sky & Telescope* but yet a nice compliment to it, *Astronomy* also covers the latest breakthroughs in astronomy, spaceflight, and astrophysics. The "Strange Universe" column by Bob Berman always presents a refreshing look at all things astronomy past, present, and future, in a humorous light. Stephen James O'Meara has a monthly column entitled "Secret Sky" that often highlights the beautiful nuances of the night sky. Another more recent monthly feature (since January 2013) is "Astro Sketching" by Erika Rix. She is also the co-author of *Astronomical Sketching: A Step-By-Step Introduction*. This has become more and more a very popular companion activity to visual observing for many amateurs, and her excellent work and expertise have been well received.

For those with specific interest in deep sky observing, the companion magazine to *Astronomy*, entitled *Deep Sky*, is an excellent resource. It was produced from 1982 to 1991 and contains a wealth of information on deep sky observing. Although out of print now, digital copies can be ordered from the Kalmbach Publishing Co.

A current journal that specializes in deep sky observing is *The Deep-Sky Observer*, published by The Webb Deep-Sky Society in the U.K. The journal offers incredible detail for objects in both hemispheres. It is available quarterly and has been published every year since 1968.

In Canada the magazine *Sky News* is very similar to *Sky & Telescope*, already reviewed, and the same can be said of *Australian Sky & Telescope*, as well as *Astronomy Now* in the United Kingdom. On a different note, the quarterly magazine *Astronomy Technology Today* covers just about everything you would want to know about the latest in astronomy equipment. If you do not already subscribe to one of these take a look at their websites to get a better feel for their content. Reading an astronomy periodical is another cost effective way to keep abreast of the latest in humankind's exploration of the universe.

Astronomical History

There are many works from the distant past to modern times that present an interesting coverage of the history of astronomy. A good place to start is with the *Cambridge Illustrated History of Astronomy* (1997), edited by noted astronomy historian Michael Hoskin. As you'd expect from a Cambridge University Press publication, the aesthetic quality of this book is outstanding, full of well-chosen illustrations and visual aids such as charts and teaching boxes. The book traces the shared history of astronomy among people of many different backgrounds and time periods even before written records or astronomical tools were created. It covers the contributions of the Mayans, Babylonians, Greeks, and also considers Arabic works and astronomy during medieval times.

Thousands of pages can be written on this topic, but what makes this volume unique is its ability to edit this information down to an amount that can be readily assimilated by the common reader. At the same time it does not seem to have left out any major strides in astronomy, and even those you may consider to be minor are given a place in the text, with some surprising revelations. Space is given to all of the astronomers that you would expect since the time of Galileo, like Wilhelm Struve, and his son Otto, Isaac Newton, and William Herschel. The last 100 pages or so are devoted to significant changes in astronomy during the twentieth century. Instead of being organized only by time periods, the book relates history by tracing the course of key concepts in astronomy. It also reveals how those concepts have unfolded as our understanding has grown and improved instruments have become available. With such great attention to detail and presentation, this volume is well worth reading.

History buffs will also enjoy the classic book by Richard Hinckley Allen, *Star Names—Their Lore and Meaning* (1963, but still in print through Dover Publications). The book presents a solid history that recounts the observations of the stars by past civilizations. Much of its focus centers around how the stars acquired their initial names and why many of those names continue to be used in our modern day. Organized by constellations, it also serves as a quick reference to the mythological stories behind the stars.

For those who may desire a much lighter history, an excellent choice is the 133-page book *Eyes On The Skies* (2009) by Lars Lindberg Christensen, who in Europe heads outreach and education for the NASA/ESA Hubble Space Telescope, and prolific astronomy writer Govert Schilling. This is an impeccably illustrated book that traces the history of visual astronomy with the aid of telescopes from Galileo to our present day. The volume reads easily and packs a lot of valuable historical details on this topic. It also cleverly interweaves practical tidbits that can help you in visual observing. Produced in conjunction with the 2009 International Year of Astronomy, this book deserves a place on the shelf of every amateur astronomer.

Biographies of Significant Astronomers

The well-researched volume *The Immortal Fire Within* (1995) by William Sheehan is an excellent biography on the life of E. E. Barnard. Most of its chapters are short, providing brief bursts of information that beckon you to continue reading. This is the fascinating story of a late nineteenth century youth with very little formal education, who like the phoenix rises from the ashes to become a world renowned observer and professional astronomer. It does an excellent job of centering on his character and leading the reader seamlessly through the account of how it drove and shaped his development in astronomy. This book is sure to keep you inspired to persevere in your own pursuit of visual astronomy long after you finish it.

In *Clyde Tombaugh—Discoverer of Planet Pluto* (2006), author and renowned observer David H. Levy recounts the life of the man who discovered what used to be referred to as Planet X. Although an image of Pluto was recorded in 1915 it went unnoticed, and the mystery was not solved until 1930 when Tombaugh (1906–1997) also imaged the faint interloper. Levy's respect and admiration for Clyde Tombaugh shines through brilliantly in this account, and more importantly so does Tombaugh's commitment and passion for astronomy. A voracious reader, he taught himself about the stars and telescope making. It was not until after his famous discovery that he went on to the University of Kansas as a recipient of the Edwin Emory Slosson scholarship. The biography does not stop with his early years, but continues to take us through his entire life, which was dedicated to helping others.

Although many biographies of Galileo have been offered over the years, the 2010 publishing of *Galileo—Watcher of the Skies* by David Wootton stands at the top of the list. In the introduction we encounter the engaging story of how much of what we know about him was almost lost to the trash forever. Many valuable papers with elements of his life have been lost, and against this background we can move forward, realizing that Galileo's history (like most figures in the distant past) cannot avoid being shrouded in some mystery. The author successfully offers not only a well-researched presentation of the facts, but he also uses appropriate questions and prose to guide the reader through the things that are unknown and how knowledge of those facts might have affected our view of Galileo in history.

Astrophysics

This is quite a broad topic that can be subdivided into many different areas. Reading more on astrophysics enhances the comprehension of what we see at the eyepiece and makes it so much more pleasurable as we contemplate the awesome nature of objects! You may wish to explore galaxy formation, stellar spectra, black holes, or the origins and nature of our Solar System, to name just a few areas. Let's begin with a very attractive and succinct publication that delves into the nature of stars.

In his book *The Hundred Greatest Stars* by James B. Kaler (Springer, 2002), we find a reference for the reader who wants to understand more about stars than just their color, but does not have time to read a large 500-page work. Here is where this title comes to the rescue. The book is dedicated "To the community of amateur astronomers," right away giving us the sense that he understands his audience and has tried to meet their needs. In just about 200 pages he successfully turns what can be a very complex subject into an entertaining read.

He begins by explaining that he has collected information on his favorite stars based on his experience as both an amateur lover of the stars and as a professional (he is Professor Emeritus of Astronomy, University of Illinois). A perusal of the contents quickly confirms these selections are no doubt among the favorites (or at least well known) of many new star gazers, such as Antares, Betelgeuse, Deneb, Mizar, Polaris, and Sirius. Others may be unfamiliar to most beginners, which significantly adds to the value of the book. Some of those stars are 51 Pegasi, Gamma Draconis, Monocerotis, Rasalgethi, T Tauri, and Tycho's Star. The book's introduction is a brief but thorough primer on the spectral classification of stars, luminosity, mass, and other properties.

Each star is covered by two pages, including a photograph of the star along with its basic data. The descriptions of the stars are beautifully written essays that express scientific information in the vernacular. This makes it an excellent tool for personal enrichment and also for sharing information with visitors at astronomy outreach events. The book concludes with a chart listing all 100 stars and their coordinates for easy location in the sky, and a glossary of key terms. For those wanting a brief tour of stars and a simple introduction to astrophysics, this title is highly recommended.

The book *Extravagant Universe* (Princeton, 2002) is an extraordinary work that recounts the 1998 discovery that the expansion of the universe is speeding up, due in large part to dark matter and dark energy. It is estimated that dark energy and dark matter make up 96 % of mass in the universe, which means that all of the visible matter known to us is composed of just 4 % of the total. The book was written by well-known author and speaker Robert Kirshner, Clowes Professor of Science at Harvard University who is also a member of the High-Z Supernova Team.

If the sound of astrophysics terminology usually makes you run for the door, don't leave the room yet. "Astrophysics is just a more forbidding synonym for astronomy," says Kirshner in *Extravagant Universe*. From the very beginning, the book sets the tone with an easygoing storytelling style and explains just how vast the universe really is and highlights how our understanding of it has changed during key moments in history. The debate of 1920 over spiral nebulae is well explained in the context of his team's recent breakthrough in understanding cosmic expansion. The book continues through highlights of Einstein's work, Arno Penzias and Robert Wilson (cosmic microwave background), along with many others up until the present day. Unlike a science textbook, this work is narrated as if it were a fast-moving eyewitness account of someone who was there every step of the way. It is also attractively and appropriately illustrated in key areas that support the intricate discussion of dark energy and our accelerating universe. If you are one of the many who have become fascinated by these topics in recent years, run to your local library or bookstore for this title.

Now let's turn our attention to one of many works that discuss theoretical astrophysics. Whether or not you agree with string theory, author and Professor of Theoretical Physics Brian Greene has done a lot to open the minds of people to the possibility that humankind may one day better understand the very complex subject of the origin of our universe. His book *The Elegant Universe* (W.W. Norton & Co., 1999) explains current ideas on string theory and seeks to engender excitement in the reader for scientific discovery. From the days of Einstein, scientists have sought a unified theory that reconciles the theory of relativity and quantum theory, essentially the one theory that explains everything in the universe.

Although the subject matter covered is much heavier than some of the other books discussed thus far, Brian Greene does an excellent job of making the topic more accessible to everyone. He has also published a beautifully illustrated book for children, great for fostering their interest in astrophysics. The book *Icarus at the Edge of Time* (Knopf, 2008), is a short but very entertaining retelling of the Greek tale of Icarus. The story examines what could happen when the event horizon of a black hole is encountered.

This is only a sampling of the books that discuss astrophysics and its many sub-topics. Although they do not directly engage us in the use of astronomical equipment or how to find and observe celestial objects, they still offer a valuable contribution to our comprehension of the cosmos. Studying astrophysics causes us to go deeper than simply the basic data of an object's coordinates, magnitude, and description. It helps us to grasp the awesome nature of the things we view in our telescopes, aiding us in our quest for answers to such questions as, "What is that star's temperature?" "Where is that nebula located in relation to our Solar System?" "When does the companion star in this binary system complete its orbit?" "How long has this elliptical galaxy been in existence?" Perhaps the most intriguing question that we can unravel by studying astrophysics is *why*. "Why are so many globular clusters clumped together near the center of our galaxy like bees in a hive?" Start finding the answers that will enhance your observing sessions by picking up a good selection of materials in astrophysics.

Observing Guides

These practical astronomy books can be written in many different styles, and you are sure to find one that suits your needs. A good astronomy observing guide should give more than a roadmap to various celestial wonders; it should also in effect become a navigator. The observing guide should not just list an object's coordinates but give some sense of how to locate the object visually. Each guide sets out to accomplish this to varying degrees. Some delve into immense detail about each celestial wonder listed, and others give a more abbreviated account for each object. A very experienced observer may prefer to have a more succinct summary, whereas a less experienced one may need a more step-by-step approach.

No matter what your level of experience, sometimes it is just enjoyable to sit down and read a guide with all the details, unabridged, complete with photos and charts, in effect transporting us to the sky when we are unable to look through the eyepiece.

Author Stephen James O'Meara gives readers an excellent voyage through 109 objects in his classic *Deep Sky Companions: Hidden Treasures* (Cambridge University Press, 2007). None of the objects covered by O'Meara in this book is on the Messier list; instead these are other objects that were uncovered before 1782. He describes himself as a "romantic" observer, savoring the subtleties of each object. Each object is introduced with a photograph and basic data that includes the original discoverer. Next, each one is discussed at length, including historical details, observation notes, and recent astrophysical information. A finder chart is included for every target.

Another selection by O'Meara is his *Deep Sky Companions: The Messier Objects* (Cambridge University Press, 1998). The book begins with a brief history of Charles Messier and his catalog, written by David H. Levy. Then O'Meara offers practical observing tips for enjoying the Messier objects. In both of these guides, his custom is to present his observations from the use of a 4-in. Genesis refractor, emphasizing that all of these targets are accessible with very modest instruments. Also, each Messier object is sketched, giving his visual impressions. O'Meara has keen eyes, often able to ferret out nuances in the eyepiece that often go unnoticed by others. Both of these guides are well researched and a welcome addition to any-one's collection of observing guides.

Written by Michael Bakich, *1,001 Celestial Wonders to See Before You Die* (Springer, 2010) is the guide to have if you want to have access to a wider selection of objects highlighted in just one single volume. Bakich, a Senior Editor of *Astronomy* magazine and formerly the Planetarium Director of the Kansas City Museum, is known for having a vast working knowledge of the sky. One unique aspect that clearly shines through in this guide is that the universe offers an endless number of phenomena that can be observed by an amateur astronomer. The author effortlessly takes us on a tour of his favorite 1,001 objects. The scope of this guide is incredible, ranging from the entire Messier and Caldwell catalog objects, to a generous and diverse selection from the NGC and IC catalogs.

Not all of the objects covered in this guide include a photograph, simply because of the sheer size of the list. However, the photographs that do accompany the objects are among some of the best astrophotography work that has been produced. Although it describes hundreds of objects, this guide is not just a list of data. It provides excellent observing notes with telescopes of different apertures, giving you a good idea of what to expect at the eyepiece. The book is easy to follow, as it is organized by objects that are visible during every month of the year.

A great observing guide for both beginners and more experienced observers is the older classic, *Burnham's Celestial Handbook* by Robert Burnham, Jr. (Dover Publications, 1978). Burnham was a tenacious observer who co-discovered comet 1957-f with his 6-in. reflector, and later worked as an astronomer at Lowell Observatory in Arizona. In each of the three volumes of this valuable work both the expertise and passion of the author clearly shines through. Even though some of the astrophysical data has become out of date, due to much advancement in this area over the years, the guide is still worth its weight in gold for the value of the observational descriptions. In the first volume he also gives a wealth of information that is useful to beginners as they enter an enriching activity that is full of acronyms and unfamiliar terminology. A generous number of photographs are provided from various observatories. This series has so withstood the test of time that it is still readily available for purchase after all these years.

The *Observer's Handbook*, published annually since 1907 by the Royal Astronomical Society of Canada, is an excellent general observer's guide offering a wide array of information. The book provides a month by month look at what is happening in the night sky. It also abounds with information contributed by well-known and highly experienced observers. The articles, tables, and other pertinent data goes beyond deep sky objects and also covers the Sun, Moon, planets, minor planets, meteors, comets, variable stars, eclipses, and more. It is a very valuable tool that helps observers to be prepared for significant astronomical events and improve their overall observing skills.

Helpful Things to Consider

Where can I find objective reviews of astronomical publications? Does a new title I am considering have content that is clear and concise? Does the book explain concepts well?

What are some categories of astronomy books that I should consider? How can reading a variety of astronomical publications make me a more successful visual observer?

No one person can be the sole expert on everything that is in the observable night sky. Astronomy is a collaborative effort that stretches across the entire globe, and through the eons of time. This is what makes astronomy such a great lifetime pursuit in which you can participate. You have the opportunity to both learn from the experience of other amateur and professional astronomers, as well as make contributions of your own. Current magazines are a great way to keep up to date on the latest advancements in astronomy. Try examining books from different authors of varying backgrounds to gain a broader perspective and to learn the finer details in astronomy that can give you that extra edge.

Reading astronomical publications can give you instant access to the valuable repositories of human knowledge and expertise. Understanding historical and background information helps put your own observations in the right perspective and context. It can give you the feeling of connectivity to astronomers of times past such as Charles Messier and Caroline Herschel. Modern observing guides can help you to locate targets and know what to expect at the eyepiece. Whether you will read an astronomical publication for your own enjoyment, or to seek motivation to add your own original observations and discoveries to the annals of astronomical history, take advantage of resources that help you make the best choices in reading. Of course, as in any genre, there is an innumerable amount of literature that can be accessed in astronomy. By using some of the ideas presented here you can narrow down that list into a more practical one that suits you personally. This will help you read more efficiently and maximize the time spent, giving greater opportunity for the act of visual observing itself.

Chapter 9

Astronomy Resources Online

Imagine that you desire to have an online tool that gives you instant access to hundreds of high-resolution images taken by a space telescope along with detailed new releases for many of those images. What else would you need in this tool? Perhaps a video library and a gallery that allows you to print out those high resolution photos. A section for educators and parents with high quality resources to use in schools, planetariums, and other science centers would be great. However, it would not be complete without covering the latest news and discoveries made with the telescope, including an archive of all releases ever made. Also ideal for this online tool would be an area allowing you to explore various facets of astronomy through podcasts, live video chats, articles, games, and much more.

When the Hubble Space Telescope was first launched over 25 years ago, it would have been difficult to envision this with the limited access, use and capabilities of the Internet at that time. Today, however, this and much more has become an exciting reality!

In the last two decades the Internet has quickly become a highly interactive tool providing access to knowledge on any subject. A quick search for "astronomy" in the search engine of choice will yield millions of results. There are websites and apps for the young and old, which cover everything from the raw science of astronomy to imaging and other aspects of amateur astronomy. There has also been a surge in access to social media. In a relatively short period of time we can exchange ideas, questions, and experiences with many people simultaneously. One benefit is that we are able to get detailed answers to questions very rapidly. For example we may wonder what are the top ten objects to view this month and receive lists from the perspective of many different people. We may also receive positive comments that could help motivate us to get out under the stars. Are you wondering what other

© Springer International Publishing Switzerland 2015
D. A. Jenkins, *First Light and Beyond*, The Patrick Moore Practical Astronomy Series,
DOI 10.1007/978-3-319-18851-5_9

people think about a certain telescope accessory? Chances are that there are multiple reviews and comments about it somewhere online.

Additionally, the web contains many personal blogs with experiences and advice for amateur astronomy. There are also countless chat rooms and forums where members comment back and forth about many different topics of interest. However, unlike a star party, where more emphasis is usually placed on observation and camaraderie, some forums can get bogged down by an endless variety of viewpoints and sometimes debates (although some time spent on forums can be very positive, and this will be considered later).

Although it is healthy to review varying viewpoints on a topic, at times this can be counterproductive within a chat room or forum. This is the case because of the time involved and the confusion that can sometimes arise from so many different opinions. It would also be prudent to keep in mind that not necessarily all advice for amateur astronomy online is accurate. Furthermore, some suggestions and ideas may indeed be correct, but may not be right for *you*. Everything has to be examined in the light of your own circumstances, your equipment, budget, observing location, physical stamina, eyesight, and even personal preferences.

How can we best sort out all of these references and accurately identify the most credible sources of information? As with anything, the user should beware of misinformation. Astronomy information located on the web does not in of itself equate to being accurate. First we need to look at who published the website. Is it run by a college or university? Is the content updated by a government entity or organization? Is it published by a recognized astronomical organization or is it someone's personal website?

Ask yourself: Who is the author and what is his or her level of experience on this topic? You can readily find this information by going to the "About" (or similar) page. Look for links to additional information or related links and sites. Do authorities in astronomy (and perhaps their websites) refer to or validate what this author is saying? What sources if any are listed for the material on the site—are they authoritative sources? A personal site is not necessarily a bad thing, but if the information sounds off track, it just might be. Be sure to corroborate the information by going to another trusted website (many are listed later in this chapter). A second and equally important quality to look for in an astronomy website would be its practicality.

Assessing a Website's Practicality

Once you establish with a reasonable amount of certainty that a site is credible, you still need to assess its practicality. Here are a few qualities to look for that will help you do so:

- The site should be well organized. This is not to be confused with a large site with massive amounts of information. Does the site leave you surfing for hours aimlessly through its pages? Then it is possible that it is not well organized.

- Visual appeal is also important. This means that there should be a good balance between text, images, and visual (such as illustrations, tables, or videos) instruction. A website that strikes the right balance between these factors is visually appealing, and thus will help you to retain the information.
- An effective site should be easy to navigate. Are there clearly designated tabs on the website or is it just one large page containing a continuous flow of information? These tabs will allow you to quickly find what you need. On many sites there are also subpages that are well defined and a site map is also another extremely helpful feature to help you navigate around.
- Another mark of a first-class astronomy website is a search box where you can locate information by entering in a key word or phrase. This will immediately list for you all parts of the site that touch on those key ideas.
- Is there a contact and/or feedback area? Creators of a site that are serious about keeping it up to date want to hear from users, as they will be the first to notice links that have stopped working or other similar malfunctions. They will want to hear feedback both positive and constructive regarding the functionality of the website and things that could be improved. At times you may have questions about data, downloads, and so forth. A contact section that is well monitored can be a lifesaver when you need clarification or information that is not contained within the site.
- Look for a site that gets results! A results-oriented site that has practical benefits is best. More than containing immense amounts of information, does it have useful files that can be downloaded (for example—videos, step-by-step instructions, observing lists, etc.)? Does it provide links to places where you can receive more instruction? Is there an app available for your mobile device? In other words, the best astronomy sites seem to anticipate your needs and answer your questions before you even formulate them in your mind.

Internet Safety

At this point a brief reminder about safe Internet searches for your children would be advisable. Even the most innocent terms entered into a search engine can sometimes inadvertently pull up undesired results. It is always a good idea to supervise children when they search the web. In cases where you cannot oversee their activity, most browsers and search engines have a setting that prevents undesirable search results. Also be aware that online forums or chat rooms can also be used to exploit the good intentions of children; therefore some parents choose to restrict their child's use of them so that it is only in their presence. Astronomy can be great fun when shared with other people who have similar enthusiasm for the night sky. However, it is important that a child not be left to arrange in-person interactions on their own. As you know, people online are not always who they say they are, so be sure to monitor who your child has been communicating with over the Internet. Additional suggestions regarding Internet safety can be found on the National Children's Advocacy Center at the following web address: http://www.nationalcac.org/prevention/Internet-safety-kids.html.

Let's continue our discussion by examining several great websites. First, many governments provide public access to a wealth of scientific knowledge acquired from organizations that are funded by the government. One example is the site: www.science.gov. This site provides public access to U.S. federally funded science research. It is maintained by CENDI (Commerce, Energy, NASA, Defense Information Managers Group), which represents various science agencies and national libraries. The website is the result of cooperation between 19 U.S. science organizations and is a gateway to over 2,200 scientific websites.

The science.gov website is a reliable resource that points the user towards several great astronomy sites. By going to the index page and selecting "Astronomy," dozens of references appear from various authoritative sources such as NASA or the National Science Foundation. For example, after clicking on "Astronomy" there is a link called "Ask a High-Energy Astronomer." This takes you to a site maintained by a small group of volunteers from NASA, and on that site is a feature called, "Ask an Astrophysicist." The site contains an archive of past questions that were submitted to an astrophysicist. If the answer has not already been covered in the library archive, and is within the realm of expertise of those that support the site, you may submit a new question that will be answered via email.

The search feature on the science.gov website is also another very powerful tool that pulls information from credible sources and organizes the search results in a list that denotes whether each entry is from a science.gov site or not. These results are then subdivided into three categories: text, multimedia, and data. Each of the references given on the topic entered can be sorted by rank, date, title, or author. Registering with the site will allow the user to set up alerts on new information available for specific topics of interest. Although there is an immense amount of data available through this gateway, it is very well organized and relatively easy to navigate.

In the United Kingdom, the publicly funded Science and Technology Facilities Council also maintains an excellent website (www.stfc.ac.uk). Look under the Research tab and then select "Astronomy & Space Science" for access to information on a fantastic range of astronomy topics such as double stars, dark matter, submillimeter-wave astronomy, the James Webb Space Telescope, Mars, and much more. The website is well formatted for the general public to easily navigate through to important points of reference. Among other things "The News, Events, and Publications" tab lists current events that are open to the public related to astronomy and other branches of science. The "Public Engagement" tab is very well done and provides a wealth of information via articles, videos, and downloadable features.

Want to learn how particle physics, computers, large telescopes, and preserving our dark skies all work together to enhance our understanding of the universe? That and much more can be found here under the sub tab "Explore Our Science." Within this section is a sub-tab called "Dark Sky Discovery" (this area can also be accessed directly by going to www.darkskydiscovery.org.uk), containing a lot of basic tips for beginning stargazers. There are even basic pocket star charts that can be downloaded and printed out for use under the night sky. This would be a great item to print out for children or to pass out at public star parties to help people recognize some of the constellations. Another interesting feature is a map of many of the United Kingdom's dark sky sites with details and directions to these locations. All of the locations are rated as to the quality of the dark skies, and notable astronomy events are also given for each site. This part of the website focuses a lot on providing resources for various groups and educators and is well worth a closer examination.

HubbleSite (www.hubblesite.org), produced by the Space Telescope Science Institute, can be considered the premier astronomy website. It is not simply a site about the Hubble Telescope and its imagery. The astronomical insights that you will find on HubbleSite extend far beyond a collection of images. This site possesses all of the six key attributes that we discussed earlier. Let's review this website in light of those six features. First, HubbleSite is well organized. At the top of the home page are eight main tabs that will help you to explore this website. They are "NewsCenter," "Gallery," "Videos," "Hubble Discoveries," "Hubble Telescope," "Explore Astronomy," "Education & Museums," and "The Future: Webb Telescope." Along the bottom of the home page are three sections containing large icons to showcase their features. The connect section lists other ways to follow news from Hubble, such as Facebook, Google+, and Flickr. The next section provides links to all of the podcasts promoted on the site. The third section highlights services such as an email notification of latest news, several blogs, and YouTube.

You are sure to be impressed by the visual appeal of the site as well. Far from simply listing items with links, each of the features mentioned above has been carefully paired with an appropriate visual aid so as to make it easy to decide what you will explore next. As you can well imagine, there is no shortage of stunning astronomical images on HubbleSite. However, the site goes much further by providing many great videos that are educational and also entertaining. There are so many in fact that you may wonder just which one to try first. One suggestion would be to try the "Tonight's Sky" video podcast in the "Explore Astronomy" section. This brief video will give you a taste of the high quality that permeates the work on this site.

The site is very easy to navigate. When you enter one of the main tab headings from the top of the home page the screen changes to a pre-assigned color and the title heading always remains visible to ensure that you will not become lost within the site. Essentially everything you see is a hot link to more information that is always visually appealing and often contains sound and video as well. You will find that most if not all features are clearly marked and explained so you will know just what will happen before you click somewhere.

The fifth attribute that makes a successful astronomy website is a powerful search tool. Wherever you are on the HubbleSite, you will always find the "Search all of HubbleSite" box at the very top of every page. This is very practical when needing to quickly locate material on a specific object or topic. It will pull up everything the site has about the topic, including articles, new alerts, images, and videos.

The contact area is practical because it not only provides a way to email an administrator with a question or comment, but it also lists some of the more commonly asked questions, so as to in many cases alleviate the need to send an email. In view of all these features, you are sure to get good results. As if all of this were not enough, HubbleSite also provides additional downloads, links to other sites, and many other tools in the "Reference Desk." Although no website can be perfect or completely error free, you will no doubt find that this site does an excellent job of helping you to more easily and efficiently expand your knowledge of our universe and current scientific achievements.

Space Place Prime (spaceplace.nasa.gov) is an excellent website developed by NASA to foster in children an interest in science and astronomy. If you are participating in astronomy outreach presentations, this is an excellent site to pull up on your iPad or tablet. From the landing page to every other part of the site, the impression is obvious—this site is meant to motivate kids to learn about astronomy. There are pictures and videos, activities and picture books, games and puzzles. Each section answers questions that children want answers to, such as: Why are there so many craters on the Moon? How did the Solar System form? What's it like inside Jupiter? The site has a lot of cartoon-style drawings that add character and appeal to its message. It is large enough to keep kids captivated for hours. This site is highly recommended to introduce children to the universe.

StarDate is an outreach tool used by the University of Texas's McDonald Observatory to provide public education in astronomy. Since 1978 StarDate has broadcast a daily radio program that currently airs on more than 300 stations. This broadcast keeps modern listeners up to date with current objects in the night sky, along with providing historical and scientific information. Additionally, *StarDate* magazine has been in print for over 40 years. This full color, bi-monthly magazine covers various topics in the science of astronomy, the night sky, space exploration, and other pertinent aspects. If you are still relatively new to astronomy, StarDate online (www.stardate.org) will provide a wealth of information on a broad range of astronomical topics, allowing you to see which aspects of amateur astronomy interest you the most. As in all of the other sites we have considered, StarDate meets the test of excellence.

The Stardate homepage gives users quick access to some of its more well-known features, such as its current podcasts, featured videos, and stargazing tips. By clicking on the "Stargazing" tab itself, much more information becomes available. This area is subdivided into eight logical sections: constellations, Moon phases, sunrise and sunset, meteor showers, eclipses, planet viewing, beginner's guide, and events. Each of these sections not only provides information on the named topic but is actually geared toward what you can observe during the current year. In effect it is a short, concise online almanac, and it is great for picking up information on the go for yourself or for use during public outreach events. A favorite is the section on constellations, which presents some lore about the most visible ones, along with details about each constellation's prominent stars, objects, and other interesting notes.

The section on Moon phases provides an interactive Moon phase calendar, a great reference for planning your observing nights for the Moon and planets versus deep sky objects. The sunrise/sunset calculator provides times for quite an extensive list of cities all over the world. The next three sections—meteor showers, eclipses, and planet viewing—all provide dates or months throughout the year for the stated objects. They also provide answers to common questions regarding these phenomena. Finally, the beginner's guide and events sections provide very basic introductory information helpful for people completely new to stargazing. It is very well written, clear and concise, and simple enough even for older children to comprehend.

Another significant feature on the Stardate website is the "Astro Guide." Beyond its attractive image gallery and video section, it also provides detailed information on the planets and minor bodies such as comets. The guide continues to give a brief look into the formation of galaxies, and also delves into several interesting aspects of astronomy such as dark matter, dark energy, supernovae, and more. This section concludes with a brief glossary. If you are a teacher you will find some interesting lesson plans to help make the science of astronomy come alive. These can be found under the "Lessons & Activities" tab. One final note—this site gives a lot of additional information with links to other less obvious parts of the site, so be sure to look for the hot links that will lead you to more of its features.

Now let's turn our attention to the Night Sky Network (http://nightsky.jpl.nasa. gov), which is sponsored by the NASA/JPL public engagement program called PlanetQuest. This is a smaller site that's packed with helpful and concise information regarding the night sky and how to connect with other people. Among some of its main features are a night sky planner, an events calendar, an astronomy club directory, and much more. After entering your location on the "Club & Events" page, you will see a calendar listing many public astronomy events in your area.

This page will also provide a list of clubs in your area, along with links and a display of their addresses on Google Maps. Want more information about a specific club? Simply click the link and the information is provided right within the Night Sky Network's web page. The format is simple and clear, essentially like a miniature version of the club's own website. It includes a listing of contacts and locations for the club. For example a club may meet in one central location for general meetings, but have several locations for star parties and/or public outreach events. You will find all of them listed and mapped out under the specific club info that you have selected. General messages can also be sent via form to club coordinators, or a request can be submitted for an event such as a lecture or a star party for a specific group. Also, if you prefer, there is a link that will take you directly to the club website.

The "Night Sky Planner" tab provides information such as sunrise/sunset, moonrise/moonset, and a picture of what's up in the current night's sky (supplied by EarthSky.org). There is also a brief video of celestial happenings for the current month. This page also includes other tools, such as the current month's free sky map, which can be downloaded and printed for use under the stars.

This is by no means the end of all the available resources on the Night Sky Network website. By clicking on the "Outreach Resources" tab, and then the "more activities and resources link," you will find dozens of informative activities, resources, and videos for both children and adults. One very prominent resource is the monthly Universe Discovery Guide. Each guide is very well written and beautifully illustrated. For each month a few key objects are featured in a story with many interesting facts. The guide will also list the visual magnitude, distance, apparent dimensions, and actual dimensions. Finding the object in the sky is made easy with the "How to Find It" feature that includes a finder chart for the area, and a complete all-sky chart is also included at the end of each issue. There are several activities inside, along with a section entitled, "How Do We Know." This is a section that answers astrophysical questions in an interactive format, containing links to NASA science sites and videos. All of the guides can be downloaded as PDF files in full color or in red to protect night vision—a great way to learn a little bit about each constellation.

Sky & Telescope magazine's (www.skyandtelescope.com) website is replete with practical information and covers many aspects of amateur astronomy for the beginner. Its online observing guide is very extensive and practical, with articles ranging from suggested objects to view in every category to choosing astronomy equipment. Other observing tools include several apps for your smartphone along with online interactive star charts, and podcasts to keep you up to date. Look under the "News" tab for the latest research in topics such as observing news, black holes, exoplanets, stellar science, galaxies, the Solar System, cosmology, professional telescopes, space missions, and more.

The magazine also has a rich history of articles featuring ideas from amateur telescope makers around the world. Some of the major ideas have been included online. Are you unsure if a particular product is right for you? There is a very extensive list of videos featuring the products of telescope and/or eyepiece manufacturers. Some of the companies represented are Meade, Celestron, Apogee, Explore Scientific, and TeleVue. Among the most informative and enjoyable videos to watch are those featuring Al Nagler, founder of TeleVue and Scott Roberts, president of Explore Scientific.

Be sure to also visit the websites that accompany the magazines *Astronomy* and *Astronomy Now* (UK), which are found at www.astronomy.com and www.astronomynow.com, respectively. The content on the *Astronomy Now* site focuses on readers in the UK and is much more streamlined. It also features many news items and videos related to current spaceflight missions. For Canadians, the *SkyNews* magazine and its accompanying website (www.skynews.ca) is excellent for finding local resources and groups. Whether or not you live in Canada, this site still has excellent articles to help the beginner become more adept at using equipment and viewing the universe. *Astronomy* has a much more extensive photo gallery online, where hundreds of user photographs are posted. You are also sure to benefit from the excellent video library that features a seasonal coverage of selected objects visible in small and large telescopes. Each of these websites is similar to the website previously described in detail, but each has their own strengths. It may simply be a matter of personal preference as to the one with which you will spend more time.

Finding Encouragement in an Online Forum

We live in a time when we have easy access to communication with people from all over the world. Interest in astronomy is global, and so it makes sense to take advantage of all the ideas and encouragement we can find. Even your local astronomy club may provide a way to communicate with others from your group through a forum on their website. Others use a server email system that automatically sends comments or questions to all members. There are also an endless variety of other social networking sites that are configured to serve a similar function. One benefit is that it helps us to feel a part of a community of people that have a passion for the universe. It also gives us a place to find encouragement when we delve into new aspects of astronomy. In the parts of amateur astronomy that are away from the telescope, an online partner may be all you need.

Ever want to try sketching objects at the telescope? There are many others who are very skilled at sketching and perhaps not so long ago were just like you—looking for advice and are now ready to share their experience with you. Sharing tips and sketches with each other can help confirm whether or not you are on the right track. Also, before making purchases you may have a question about a piece of equipment and how it might perform for your particular needs. Chances are someone is out there who already has used the product and is willing to share their insights. Ever in need of an observing partner? A quick post on your local club's forum may be just what's needed to get someone away from the television and out into the field to observe with you. Remember, even Charles Messier worked closely with his friend Pierre Méchain, who discovered and recorded several objects included in the Messier Catalog that Messier did not initially find.

Where is a good place to start? Begin by asking others in your local astronomy club. If you are not yet a part of a club this would be a great time to get acquainted with other people who share your interest. You can also check out amateur astronomy magazines or their websites. From time to time they often refer to good online astronomy communities that you may wish to approach. One such community is offered at Cloudy Nights (www.cloudynights.com). Cloudy Nights provides many articles on various aspects of amateur astronomy, a forum for reviews of all kinds of equipment, and a place to find notifications of various events. Here are a few of the regular topics found in its forums: general astronomy, equipment, astrophotography and sketching, observing (grouped by type), and specialty (miscellaneous). Here you will find people at all different levels of experience in astronomy and at least one person interested in the same or similar parts of amateur astronomy as you.

Take as an example the forum on deep sky observing. Although it would be nice to always have an observing buddy outside at the telescope with you, that is not always possible. At times, observing is something that we may have to enjoy alone for whatever reason. However, the experience can still be shared with an observing partner online. This is especially valuable when you are unsure if you spotted the correct object, when that voice inside of you says, "Was that *really* the Ghost of Jupiter that I saw or did I see something else?" Those are the times that you can compare experiences with another observer in the forum. Be sure to note down your experience in detail (and sketch your observations if possible) during your session, or very soon thereafter, so you will not forget what you saw when posting comments online.

Helpful Things to Consider

What astronomy websites or topics interest me the most? Are the websites that deal with those subjects easy to navigate and visually appealing? When I participate in forums, is my experience an informative and encouraging one? If not, could I find someone from a local club to become my online observing partner?

Are the astronomy websites that I use effective, practical, and the kind that lead me to my desired results?

Here is a list of additional websites that may benefit you:

International Astronomical Union	www.iau.org
The Astronomical League	www.astroleague.org
Association of Lunar & Planetary Observers	www.alpo-astronomy.org
International Dark-Sky Association	www.darksky.org
Spitzer Space Telescope	www.spitzer.caltech.edu
Chandra X-Ray Observatory	www.chandra.harvard.edu
AAVSO	www.aavso.org
Worldwide Telescope	www.worldwidetelescope.org
Royal Astronomical Society of Canada	www.rasc.ca
Royal Astronomical Society (UK)	www.ras.org.uk
European Southern Observatory	www.eso.org
Jodrell Bank Discovery Centre	www.jodrellbank.net

Whatever your preferred manner of involvement, be it a social networking site, a video chat, a forum, a club intranet, or simply an email exchange with someone, try to give as much as or more than you take in. Other people will appreciate *your* thoughts, opinions, advice, and encouragement just as much as you do theirs. Having this reciprocal mindset will help ensure that we continue to open the minds of more curious persons to the awesome world of amateur astronomy.

The many resources on the Internet are your tools, and they should not dominate or unduly possess your time. Whatever is your preferred level of involvement online, be it social networking sites, emails, or something more, allow its use to help you enjoy the beauty of visually observing the sky for yourself. Furthermore, as will we see in more detail in the next chapter, being able to share that joy with other people who have not yet seen some of these beautiful celestial sights is also very rewarding. The Internet is a marvelous tool, bringing vast amounts of information to our fingertips at lightning speed. This savings in time equals more time spent observing and in turn sharing the universe with other people. It will also help you to continue making a success of astronomical observing.

Chapter 10

Inspiring Others

In many countries, people gather in large numbers to view what they believe to be a spectacular show. Long lines of people form hours before the show begins as they wait in eager anticipation to see every frame from the opening scene to the finale. Sometimes tickets are required, either free of charge or for a nominal charge. There are ushers and guides ready to share the exciting phenomena with everyone, until each individual has had his or her own personal view of the show. Oohs and aahs can be heard as spectators utter sounds of surprise at the enormity, beauty, or intricacy of the wonders that they see. These are not the long lines that fill movie theaters or similar entertainment venues. Instead, they are the growing number of people gathering to see the heritage that belongs to every person on Earth, the privilege to see the wonders underneath a truly dark sky.

More and more people of all backgrounds are discovering every day that the universe is a place that everyone can learn to observe and understand. Instead of relying solely on pictures they are going outside, with the help of experienced people and all sorts of optical instruments, and they are looking directly at the night sky for themselves. Once this beautiful heritage has been discovered, it begs to be passed on to even more people, so that the quest for understanding the cosmos will never end.

So then, how can we pass on this sense of wonder to others who have not yet experienced it? How can we get them into the line in the first place, to wait for their opportunity to see firsthand just how awesome astronomy really is?

The ways in which we can inspire others to become excited about astronomy generally come in one of two forms—either a casual approach or a more structured one. The casual approach is informal and does not require much preparation at all. The structured approach involves much more thought, preparation, and follow up,

© Springer International Publishing Switzerland 2015
D. A. Jenkins, *First Light and Beyond*, The Patrick Moore Practical Astronomy Series,
DOI 10.1007/978-3-319-18851-5_10

but the results are more impactful and last much longer. Most of this chapter is dedicated to discussing those types of motivation. However, let's be clear—structured does not mean boring, and neither does it insist on a strict lecture format. Really, although somewhat structured it should still be fun; otherwise people will not likely be inclined to listen at all.

Casual Approach

Let's first briefly consider some of the casual methods of motivation. This includes just about anything that is simple and unprepared, but is still welcoming. For example you may be attending a parent-teacher night at a school or a similar function where parents can exchange ideas. This is a good chance to get kids involved in fun and educational activities. You could mention what you're doing over the weekend, such as looking at the lunar eclipse, Uranus, or the Andromeda Galaxy, etc. Many people react very positively when they hear someone is interested in the stars and just never knew how to get involved in something like that. So at that point you invite them to an astronomy club meeting or a public star party. Easy as pie right? It involves no real preparation, just a friendly conversation that leads to someone expressing some level of interest in what you are saying and finally to an invitation to learn more.

Internet access is now common and widespread. Even children can get quick access either with a smartphone, tablet, or at school. A great astronomy app for children age 8 and older is the *Space Scoop* astronomy app available on Android. It uses simple language, lots of color imagery, and strives to present astronomical material in a way that is fun and updated for the twenty-first century. Another one now available is the *Space Place Prime* app by NASA, a spinoff of its website (spaceplace.nasa.gov). It offers great astronomy images, articles, videos and games for kids. It is available on both Android and iPhone or iPad (Fig. 10.1).

Another example of a casual approach to motivation would be referring to the Internet in passing. For example your child has some friends over and the subject of what to study in college comes up. You start to talk about opportunities within various STEM fields (much to your child's dislike, since he or she has heard this a dozen times before) and show him or her a few websites that feature current students of astronomy at major universities. By the way, if you have not looked for something like this in a while, you should. Universities know how hard it is to attract students to this path of study; it takes years and needs to start early (as young as possible and by high school at the latest). Take a brief look at Cornell University's "Current Grad Student Spotlight" pages. It does not simply explain what a career in astronomy is about, it lets the current students describe their own experience in this pursuit (see http://astro.cornell.edu/student-spotlight.html).

On a lighter note it could simply be the inclusion of astronomy-speak in jokes or conversations. This would also include making reference to astronomy when using social media, email, and other types of communication. Also, when it comes to gift giving, what better way to provide long-lasting inspiration than to give a gift related to astronomy? Some ideas are a science or astronomy magazine

Fig. 10.1 The Space Scoop Android app, developed by Universe Awareness to share the most exciting new astronomical discoveries with children. The app is available for free download at the Google Play store. (Image courtesy of EU Unawe/ESO)

subscription, a book, a smart phone app, a pair of binoculars, or the most obvious—a telescope. The Internet is full of companies that make all sorts of astronomical gifts. Gift ideas range from the Milky Way printed on ties and constellation playing cards to home planetariums and astronomy software. With a little ingenuity and creativity, the possibilities are endless when giving the gift of astronomy. (For some great gift ideas, check out the online store of the Astronomical Society of the Pacific at http://www.astrosociety.org/astroshop/).

Reactive Outreach

When we receive a request to provide an outreach activity for a group, it can be either an open request or a closed one. In other words the person or group asking for the assistance may be (in the first case) open to any ideas we put forth that are commonly used for a particular audience (grade school children, for example). This allows us more flexibility in the presentation or activities. Pre-planned activities and events work well in this case, and it would be a good idea to have several different ones created to be ready at a moment's notice.

A closed request, however, is one that is made with a specific set of requirements and it is preferred that we not deviate from those criteria. The need is unplanned from our perspective, and at times is an urgent need. Often there are people staffed at observatories for this very purpose. However, in the case of astronomy clubs the persons involved volunteer their time and assistance, so the immediate needs requested do not always match the availability of those responding. This can be a little more difficult if there is only one person in your club or organization that is able to perform the request. If that person is not available, an opportunity could be missed because of scheduling challenges.

Although reactive outreach activities imply or can involve certain challenges, they should be viewed as a positive. The reason for this that the request is coming from a group who already wants the assistance, which usually means there is already an interest in astronomy. There is already a vested interest in seeing the event through. We just have to become better at being ready to seize unique opportunities where interest in astronomy already exists, and then do our best to grow that interest. Of course not all reactive outreach is requested specifically for astronomy—sometimes it could be a request to fill a need for a science-based activity that is kid friendly, educational, fun, and that can be delivered at the needed date and time.

Proactive Outreach

Proactive outreach is by far the more desired conduit for outreach. It is pre-planned as to content, delivery, and schedule. Most of everyone's needs can be more comprehensively met. A preview of what will be discussed can be sent ahead of time, allowing for constructive questions to be formulated by the audience in advance. If the outreach activity involves a presentation or event that must be at the requestor's location, it is important to give adequate thought to how to create the right atmosphere in a space not normally used for that purpose. To create the right setting, certain diagrams, images, posters, and other props need to be available. An outreach event at a nearby observatory where everything is already set up has a distinct advantage in requiring less preparation and ingenuity on your part. The audience can essentially be transported to the world beyond Earth's atmosphere, so that from the moment they arrive they feel welcomed and there is a sense of wonder. For children those feelings of whimsy are important to fire their interest and motivate them to consider more involvement in astronomy. However, when we do have to provide mobile presentations, the audience can feel more comfortable in their own "backyard." We just have to do everything within reason to give them the same sense of wonder when we are bringing the presentation to them.

This is just as important for adults as it is for children. Ask most adults and they will agree that images, colors, and sounds contribute to their selection of entertainment and the way they use the Internet. Presentation also takes priority in our choice of food and clothing. In the same way you can be sure that an audience of adults will be looking for the same attractive factors in any presentation that you offer. Adults likely get enough drab technical jargon at their place of employment. In any activity in which people choose to spend their personal time, they are going to want to get the most bang for the hour, meaning it has to capture and maintain interest in an appealing way. They, much like children, want to be able to say in retrospect that they learned a lot and had fun.

We should also be proactive about returning to groups that enjoyed an initial program whether it was a star party, presentation, or some other event. People who enjoyed an event will often want to do it again because they already know it will be good. Of course that means we have a standard to maintain, but that is a good challenge to have to deal with. Really, a return to a group that we served before begins as a repeat performance of what we did last time. Not that we will use the exact same content we presented, but the *manner* in which we presented the program and how we made each attendee feel will be repeated. For example, we probably do not remember every detail of a meal we had at a restaurant a year ago, but we probably do remember how we *felt* about that meal. This is not to say that we should just wing the material and come up short on substance, but it just illustrates that our delivery is just as important as content.

There are more and more organizations that are providing help so that more professionals in many different fields can reach a wider audience. The Internet is a wonderful tool that has been of great help to facilitate those efforts. One such company is Nepris, a company based in Austin, Texas, that started up in 2014. Already it has accumulated about 1,000 teachers and professionals in 45 states. The objective is to connect people from the STEM fields as well as the arts, with classrooms across the country. It provides a way for teachers to help engage their students by providing virtual conversations with industry experts who sign up to be available to speak on their area of expertise. Educators that are members of the program have full access to sessions that are recorded, so even if their class was unavailable for the live session, they can still access the multimedia presentation. Although people who offer their time are providing a reactive response to requests, it really is proactive since they are actively listed as available to help with a specific presentation. Perhaps in the future this or other similar organizations will continue to make an impact in spreading an interest in STEM fields.

These platforms may also be a tool that astronomy organizations may wish to look into for another way to provide outreach.

Astronomy outreach is also about making good partnerships. No one person or organization by itself can reach everyone; we all must do our part. A great way to amplify the reach of your proactive motivation program is to partner with an astronomer willing to periodically give of their time. Local universities and colleges usually have at least one person with whom you could team up with. That person will have other contacts that will grow your base of regular outreach recipients and may give you access to more. Often educators in colleges are looking for someone in the community to make their areas of study more visible, so in the end everyone can benefit from these joint efforts.

The result is an audience that benefits from a wider variety of participants, access to an expert in astronomy from an academic point of view (if not also in research or in projects), increased validation and wider recognition. In turn, it could also lead to quality presentations for your astronomy club as well, which could potentially increase membership. These are just a few ways that we can stretch beyond our current limits and seek positive change in a rapidly shifting world. In order to keep up we have to get ahead of the curve. What are some ideas on what activities can be planned for outreach? Also, how can we offer effective presentations and public viewing nights?

Public Outreach

Hopefully the outreach that we provide extends beyond one event and grows into an ongoing program or at least a periodic one. This could entail public viewing nights at a club's dark sky site or observatory, star parties, or presentations at the requesting group's location. Under the outreach section of the Night Sky Network (http://nightsky.jpl.nasa.gov) there is an innumerable reservoir of resources that can help anyone to provide more than adequate outreach activities that are unique, engaging, and informative. Among many of these very useful tools are the observing cards. These have concise but interesting things you can say about objects you may present for viewing in the telescope at star parties.

What about the amount of time spent in a presentation to a group? The length of time should not be too long. A good rule of thumb is to end the presentation while the audience is still enjoying it. This will leave your audience with the desire to hear more, and so they will be more likely to invite you to return. It is that desire to hear more that creates an ongoing outreach base that you can visit on a regular or periodic basis. Keep in mind that your audience cannot hear your next presentation if they do not know about it. One simple way to be sure they do is to have small individual sign-up sheets available to be filled out as people enter or leave. This is one of the best ways to capture contact information so that you can notify them about your next event.

How should a presentation be conducted? Don't preach at your audience. Instead involve them interactively. Try to reach their mind and heart with simple ideas that have an impact. Whether a presentation is geared for a more simple, basic concept, or for a more specialized, technical topic, keep your material clear and simple. Strive to be an engaging and conversational speaker. Ask your audience easy questions, such as "How are you all today?" and "What is the largest body in our Solar System?" It always makes people feel good to be involved and creates their active engagement in your presentation. Since some people are shy, do not single anyone out for an answer, just ask the audience in general. There is bound to be at least one person who will respond. Also remember to relax, smile, tell short stories to illustrate important points, and use visual aids (whether we're 9 years old or 90 years old we love them).

Something that is usually lacking at lectures or presentations is food or light snacks. There is nothing more distracting to a listener than being hungry. Great facts about being able to see the Andromeda Galaxy 2.5 million light years away, or how to observe the phases of Venus, will not be heard by hungry listeners. What's worse is that an audience distracted by hunger will always connect seeing you with hunger. Even if you give a short thirty-minute presentation, in their minds it will seem like hours. What can be done? A budget for a full course meal is probably not reasonable, but you could provide light snacks and drinks to help. If you are providing outreach to a school, they could supply snacks for the children based on your recommendation, and this is probably the required and wise choice in that case. Otherwise, expensive snacks are not necessary. What you are looking to do is distract the minds of your audience away from hunger and onto your topic. So even candy will do the trick, and instead of connecting you to hunger, they will equate you to something they like and be better able to focus on what you have to say. Other snacks you might try include fresh fruit, granola, popcorn, small sandwiches, and cookies. Be sure to inquire if anyone in your audience has food allergies, because many people have severe allergic reactions to certain foods (such as nuts).

Do you need a question and answer session? Yes, this is a must. People already will have questions whether they appear to or not. Opening up the conclusion of your presentation for questions and answers gives an individual your approval and permission to ask whatever is on his or her mind. Many people are shy and will need this approval from you.

After the session is over, a speaker or presenter will need to be accessible for a few minutes. Why is this necessary? After such a wonderful presentation it would not be good for your audience to feel that it was canned or that you have just turned off the music now that it is over. Be friendly and accessible! This also involves being available in the days and weeks that follow. Always provide an email address so interested persons can reach out to you for more information. Also let everyone know the schedule for upcoming club events and meetings.

Looking for an outreach activity closer to home? Set up your telescope, and they will come! Just the sight of a long tube pointed to the sky is enough to turn heads. People are inquisitive, and most everyone will immediately recognize that you have a telescope, but not everyone has had the chance to look through one or ask questions about what you can see in a telescope. Seeing you (a neighbor) outside with a telescope is a proactive way to invite them to join you on a short stellar or planetary voyage.

At public events several effective displays can be a good method to offer information. These can be set up at a club location or other locations that are agreeable such as restaurants, schools, community centers, community exhibitions, and observatories. Effective displays can range from the very simple to more elaborate, and often resonate well with visitors as they can get short snippets of information that are very powerful demonstrations of key concepts. Several topics can be absorbed in a short period of time, and people can grab one idea and keep moving. Also, people only visit the displays that interest them, but after overhearing others or seeing the reactions of other people, they may visit the display they initially passed by.

Ways to Advertise Your Outreach Programs

If you are already a member of an astronomy club you have no doubt experienced at least some occasions of low attendance. That does not necessarily mean that the meetings are not informative or enjoyable. Rather, it is sometimes just an indication that not enough people know about your astronomy club and all the activities associated with it. Also, in today's world of entertainment and electronic gadgets, there is a big competition for people's time. We have to get the word out! There is something just as exciting as a smart phone or gaming console, if not more so, and it is right above our heads. So, we have monthly meetings and star parties that are scheduled, and we may even have a website that explains all of this. However, that is simply not enough. What are effective ways of drawing in more attendance and members?

As explained earlier, the Night Sky Network has created a wonderful interface that allows users to find all astronomy-related events in their area. Be sure that your club is affiliated with Night Sky Network so that people using it will be able to see your programs and activities. If you do not have a website, Night Sky Network can even help you to create one complete with a calendar of events. The advantage of being affiliated with this larger network is that there are so many wonderful tools on its site, it is bound to make an impact on persons that are seeking involvement in astronomical activities. This creates more interest in their local club where they can enjoy being around people with similar interests and concerns.

A club's own website should always be kept up to date and look attractive. If you are not sure of how the website is being perceived, ask members to vote on its attractiveness at your next meeting. Do not leave it in the hands of just one person to hold the entire responsibility to make the club website appealing. Try adding a well-made video to your club website. It is now common for people to use YouTube or other videos in their decision-making process as far as involvement in activities.

Why not make use of this on your website? Most personal computers come with basic software that can aid in compiling video footage along with photos and text. We have to make the best use of technology to offer an effective way of communicating the importance of what our astronomy clubs have to offer. Remember, people (adults and children) are using desktops, laptops, tablets, and smart phones. A great website that looks inviting on all of these devices will be a website that is regularly viewed and read. If it is read on a regular basis, then people will start showing up at more meetings.

Other than a great, up-to-date website, how else can we advertise outreach events? Try your local school district. Schools welcome these kinds of activities. They can be standalone events or held in conjunction with other events such as fund raisers. Try suggesting "Dollars For Stars." People who attend get to have a look through a telescope for the cost of $1. Every dollar contributed goes to help fund a school function or another cause. Use local community newspapers and radio to keep people up to date with the date, time, and location of your outreach programs. In some cases public radio stations may make announcements of happenings in the local community. They may even do so free of charge if it is for an activity in association with a non-profit organization.

There are many other groups that can be approached as well. Of course the Boy Scouts and Girl Scouts should be on your list. They have always welcomed astronomy-related activities as a part of their routine, and this provides a very receptive audience. In addition to public schools, try approaching private schools. It is likely that you have driven by many private schools without even thinking about them, so be sure to look carefully next time you drive past a school. Also, check the Internet for a listing of all of the private schools in your area. These institutions are very flexible as to the kinds of activities that they can schedule, and there is a shorter approval time because there isn't the red tape most public school districts require.

Another place to advertise is at your local library. Inquire about forming a local sub-group of your astronomy club that meets during the daytime at the library. This may be very appealing to older persons who may not wish to be out at night. It would also be a good place to meet for home-schooled children. Parents of children that are home schooled are always looking for ways of getting their kids out on field trips and to interact with other children their own age. The idea for both of these two latter mentioned groups is to encourage them to also be a part of the night sky viewing that your club offers.

Are there companies with a large work force in your area? Today's successful businesses know the value of encouraging unique activities outside of work. Ask permission to post an ad about star parties or club meetings on their intranet, or on the bulletin board in the break room or lobby. Promote visual astronomy as a way to reduce stress in the workplace. You could offer to do a 30-min after hours astronomy chat. Be sure to make it an attention grabber! Not to be forgotten is the power of social media. There are many in use that can be used to attract more members and recipients of astronomy outreach. A few are Google Hangouts, LinkedIn, Facebook, Yahoo groups, Pinterest, your own email contacts, and many others.

For public viewing you can use other places than your usual location for meetings or star parties. Try seeking out places where people already go such as school lots, school sports fields, parks, and grocery store lots (perhaps in the rear away from traffic). Be sure to seek permission before setting up in any of these places to be sure you are not trespassing or interrupting an owner's business activities. On that note, why not try a local restaurant? A partnership with a restaurant may be agreeable to the owner because it can attract more people, which can translate to more business. Programs that help kids stay productive are usually met with welcome arms in the community. This is already being done in some areas with good success. You may be thinking, what about all of that light pollution in those areas, how do we deal with that? Limit your viewing to the brighter planets and the Moon. For most people, this is a great opener and can be an enticement to join your club for a star party at a dark sky site. It will also give you an opportunity to discuss the effects of light pollution on the night sky.

These are just a few ideas on getting the word out. Sometimes it can feel easier if we just sit back and let people come to us, but it is much more rewarding to be actively involved, to reach out and find those persons who are likely to be interested in experiencing something new. In considering these suggestions you will probably come up with many more ideas of your own. Share them with other clubs at larger gatherings such as Astronomy Day. With everyone pitching in with just a little bit of help, you can stretch beyond your current boundaries, reach a much wider audience, and have a positive effect on the future of astronomy.

Start an Astronomy Club!

It is quite possible that you live in an area that does not already have an astronomy club, or the nearest one may be located too far to be able to fully participate. This may give you an opportunity to start a club from the ground up. There are resources available to help you do so listed on the Astronomical League's website (astroleague.org). Also there are suggestions on the duties of the officers in a club as well as how to handle membership, non-profit organization status, dealing with conflict, and other essential concepts.

Here are some initial things you may want to give thought to. Where will the club meet and how often? If it only consists of your family, friends, and a few neighbors, someone's home may do, or even a public park or coffee shop. Once attendance begins to swell, your local library may offer a conference room that can be used for this purpose. For larger groups a classroom, perhaps on a college campus, would offer a great atmosphere. That would lend itself to forming a partnership with science professors at the college and make it convenient for a guest astronomy speaker to come from time to time.

At first you may want meetings to be on a quarterly basis until membership becomes substantial. Then it can be increased to monthly. Since observing will probably be central to your club; the observing sessions should be at least monthly from the beginning. This leads to another important thing to consider. How can you acquire access to an observing site? Again, schools and parks (if open after dusk) are a good place to start. If you are providing outreach efforts to students at a school you would likely get permission to use a certain area for observing. Other clubs have either purchased land or have received a donation of land. Perhaps a member lives in an area where the sky is dark. Perhaps that member would not mind the club meeting on their property once a month for a star party. The key here is to find suitable areas and then be creative and gracious in your way of approaching the subject.

Where can you find members? Start with friends and family, use social media, schools, and also begin with people in your neighborhood. Does your area have a community garage sale? You could set up an astronomy display during that week-end and promote membership in your new astronomy club. When offering views through your telescope in your neighborhood or local parks be sure to mention membership in your club. Design a simple flyer to have on hand whenever you speak with someone about astronomy or during outreach events.

How will your club be organized? It need not be very complex. Keep it simple and be sure the focus of the club is to help people have fun observing the night sky. As people join, try to match the strength of the individual with any role that they might assume. In addition to the usual roles of president, vice president, sec-retary, and treasurer, there are several other important positions. These include someone to head new membership and member services, an Astronomy League correspondent, someone to organize club outreach events, and someone who will create a newsletter and website. There are a lot of benefits available to a club that joins the Astronomical League. Let's briefly discuss how that organization can prove to be helpful.

The Astronomical League

The Astronomical League is a dynamic amateur astronomy organization composed of over 240 member clubs and societies across the United States. Its mission is to promote the science of astronomy. The league is essentially the voice of amateur astronomers and offers several ways for members to connect with other amateur astronomers, to benefit from their knowledge and experience. This is done by annual meetings on both a national level and a regional level. These gatherings offer members a chance to observe or become involved in the league's annual business meetings, but they offer much more. The annual gatherings feature interesting talks by professional and amateur astronomers, exhibits, displays, workshops, and some-times also include star parties.

Each member of the league receives a quarterly newsletter called the *Reflector.* It is designed by amateur astronomers, and contains interesting articles on important ideas in amateur astronomy. The *Reflector* also keeps readers up to date with the latest in the league's activities and services. Another benefit that members have is access to the Astronomical League's Book Service. Through it members receive a 10 % discount on astronomy-related books and free shipping.

A very practical benefit of membership is access to the league's observing programs. There are almost 50 of these, and they are organized, structured activities that help people to become fully involved in all aspects of amateur astronomy. Once a program has been completed, you are awarded a certificate and pin that are presented through your local astronomy club. What's nice about this is that you know each observing program has been tried successfully by other amateurs, and there is written guidance on how to complete each one. Each program has a coordinator who can provide more information if needed. Some of the observing programs are Binocular Messier, Bright Nebulae, Double Star, Asterism Observing, Solar System Observing, Arp Peculiar Galaxies, Comet Observing, Dark Sky Advocate, and Outreach Observing. These programs offer a structured way to observe that helps keep you on track and provides an easy way to track your progress as an observer.

ALCon is the annual event sponsored by the Astronomical League that you will not want to miss. It is usually a 3- or 4-day event and has been held each year since 1939 (except during WWII from 1942–1945). Here are a few highlights from the 2014 ALCon event.

The Astronomical League President John Goss discussed an outstanding innovation originated by Marc Stowbridge of the New Hampshire Astronomical Society. The idea is to provide loaner telescopes that people can check out from their local library. These are easy to use 4.5-in. (114 mm) tabletop reflectors that are slightly modified so as to provide an immediate "out of the box" experience for the user. Some of the modifications are installing a simple red dot finder, a permanently installed zoom eyepiece (screwed into the focuser to prevent patrons from removing it), a plastic cap on the rear of the 'scope to prevent mishandling of the collimation screws, and strings that attach the dust caps to the 'scope to prevent loss. The library loaner 'scopes are then stored in large plastic bins at the library.

Once your club donates these 'scopes to the library, nothing else needs to done by the club except for any occasional maintenance that might be needed for them such as collimation and cleaning. The library handles everything after that, just as they would for books that are checked out. According to one library in new Hampshire, on average six people use each 'scope that is checked out. A 600 % return for one outreach transaction presents an incredible opportunity for astronomy clubs everywhere! For more details on this kind of program, please see the article written by John Goss that appears in *Sky & Telescope* magazine (October 2014, pp. 66–69).

Well-known amateur astronomer Larry Mitchell (Houston Astronomical Society) gave an excellent slide presentation on deep sky objects. His opening exclamation that "visual astronomy is not dead" met with applause from the audience. Larry Mitchell created the Advanced Observing Program for the Texas Star Party and encourages people to go beyond commonly known Messier and NGC objects. Listeners were riveted to his presentation as he revealed fascinating objects of all types that many amateurs have never heard of before. When they hear about certain astronomers seeing faint deep sky objects, some amateur astronomers assume that it is only possible with 'scopes that are over 20 in. However, Larry Mitchell emphasized that these deep sky objects are accessible to observers even if they use a smaller 'scope in the 8–12 in. range. You do not have to own a very large 'scope to see these fascinating objects. He encouraged everyone to increase their observing skills, to learn the everything possible about what objects are visible in the sky, and most of all to have fun whenever observing. At star parties, even when it's cloudy, you can have a great time as you meet and interact with interesting people.

Another very interesting topic entitled "Celestial Sleuthing" was presented by Dr. Don Olson from Texas State University. His discussion centered around celestial phenomena that have appeared in famous works of art. By skillful analysis of places that served as backgrounds for paintings, the precise date, time, and location can be determined for these works of art. He has traveled to many countries to unravel these mysteries in works by Van Gogh, Monet, and many others. He has also shared these celestial sleuthing adventures with his students, helping them to realize the new appreciation that comes with such precise understanding of these works that have influenced our culture. Olson's book *Celestial Sleuth: Using Astronomy to Solve Mysteries in Art, History, and Literature* (Springer, 2013) gives readers insight into many historical and literary mysteries that have been solved through astronomy.

These are just a few of more than a dozen speakers from around the country that spoke on topics of interest to amateur astronomers. Other topics included extrasolar planets, how to improve your club, planetary observing, professional astronomy, outreach, supernovae, light pollution, and many more. The annual three-day ALCon meeting is definitely well worth the effort to attend.

Let's take a few moments to get some personal comments from someone who has played a key role in the league. John Goss, newly elected president of the Astronomical League, has made astronomy outreach a central part of his work as an amateur astronomer. In an interview in late 2014, he expressed some interesting thoughts on astronomy outreach and what contributes to the success of amateur astronomy.

Author: *Tell us about when and how you first developed an interest in amateur astronomy.*

John Goss: It first started when I was about 7 years old. My big sister told me that I could see the Moon in the daytime. I didn't believe her until she showed it to me in binoculars, and from then on I was hooked. My interest really piqued during eighth grade, just before the Apollo Moon landing. Then, like many other people, I became distracted from astronomy during my college years and during the years spent raising a family. It was in my thirties that I began to really pursue astronomy again.

Author: *What are some new things on the horizon that we can look forward to in connection with the Astronomical League?*

John Goss: We currently have about 16,000 members, but not everyone is familiar with all of the things that the league is involved in. So one goal that we have is to better promote the Astronomical League, membership in the league, and amateur astronomy itself. Also given that our scope of activity is nationwide, in the future we will be coordinating more interaction with professional astronomers. Amateur astronomers will be able to get involved with more citizen science projects.

There will also be continued promotion of our observing clubs. Right now just about every aspect of amateur astronomy is being covered by the more than 40 observing programs. One very important one is the Dark Sky Advocate Award. To be awarded this, the participant must complete a series of activities related to understanding light pollution and becoming involved in offering solutions to this problem as well. For example it would include activities such as sending a letter to a business or municipality requesting an upgrade to shielded outdoor lighting, or preparing and submitting a handout for the public that details the problem along with solutions. Light pollution is a big problem, and it is up to us to vocalize it, help people understand it, and to help provide solutions. Imagine the effect that it would have if all 16,000 members of the Astronomical League would just speak to their local municipality about this issue, or even simply wrote a letter to them addressing these concerns.

Author: *What do you enjoy most about astronomy outreach?*

John Goss: There are so many different types of people who are interested in astronomy. Consider the various people that wait in line to have a look through your telescope. There are young children who probably begged their parents to bring them out. There are teenagers who may act as if they don't want to be there, but really do want to have a good time looking at the stars. There are doctors, bikers, and many other people that although different from each other, are each touched on a personal level by what they see. I enjoy seeing how all of these people appreciate these events and always walk away with something to think about.

Author: *What are some ways in which amateur astronomers can become more effective at outreach?*

John Goss: The first thing that comes to mind is that we should go where the people are. For example, if you have your telescope set up in a city park, don't park yourself so far away from where the foot traffic is that no one will want to come to you. We cannot expect people to come to us, we have to go to them. It would be better to park your telescope right next to the walkway where people are constantly walking by. Then they will be more inclined to approach us and take a look.

The second thought I have on being effective is to show the people what they want. For most people this means Jupiter, Saturn, and the Moon. This is no time for pulling out your favorite dim NGC object. The result of showing them what they want usually is that they will come back again for more.

Author*: If you could give just one piece of advice to amateur astronomers, what would it be?*

John Goss: Get outside and observe! It is so easy to say, "the Moon will be back again next month," or "Sagittarius will be back again, I'll wait and see what's there next year." Why not go out and view them now? The other piece of advice I have is to make sure that what you see is enjoyable and not a chore. Observing should always be enjoyable.

Conversation with Connie Walker

Here are some useful comments from Dr. Connie Walker, Associate Scientist and Senior Science Education Specialist at the NOAO and director of the Globe at Night campaign. Another one of Dr. Walker's passions is in astronomy education and outreach. In addition to discussing her efforts in both areas, she also offers practical advice regarding outreach and astronomy.

Author: *Tell us about outreach and what you enjoy about it.*

Dr. Walker: I enjoy working with others in the community to help educate them in astronomy and on the effects of light pollution. Creating this awareness in people is one way to effect change in behavior, with real action as its result.

I currently work as the director of the Globe at Night campaign and also sit on the Board of Directors for the International Dark-Sky Association. My work with the Dark Skies Rangers program has endeavored to help children learn how to measure the effects of light pollution, understand more about the stars, and practice the use of proper lighting.

Author: *How can more people get involved with inspiring others to love astronomy?*

Dr. Walker: Get involved with the International Dark-Sky Association (IDA) (darksky.org). Strive to understand the viewpoint of other people, not just our own desire for astronomy. An effective way people get involved is in seeing the impact on factors that may be important to them such as health, safety and security, wildlife, and energy consumption. The learning curve is really a four-step process that begins with awareness, followed by a change in our attitude, which leads to a change in behavior, and that finally hopefully results in action.

The IDA has local chapters near you, or you can start your own chapter. It provides suggestions for speaking to municipalities about concerns on light pollution and how to speak with city officials so as to pass lighting ordinances that preserve our dark skies. There is also a guide for neighborhoods, and model ordinances for municipalities to use as a whole or in part, depending on the needs of the community.

The World At Night (along with the National Optical Astronomy Observatory) hosts an annual International Earth & Sky Photo Contest to help promote awareness of light pollution issues. These photos often juxtapose the beauty of the night sky and the visible effects of light pollution. For more information on submissions to this contest, see twanight.org/contest.

Author: *Are there any initiatives that the NOAO is heading to reduce the effects of light pollution?*

Dr. Walker: Besides the educational efforts already discussed, the dark sky experts of the APSS research in Arizona (Astronomy, Planetary and Space Sciences, of which NOAO is a part) have been key proponents of lighting ordinances at public hearings and have provided educational outreach on the ill effects of light pollution for many years. Protecting our environment by limiting light pollution better serves our economy in that it facilitates job growth, reduces energy consumption, protects our health, promotes safety and security, and preserves our wildlife.

Author: *If you could give just one piece of advice to amateur astronomers, what would it be?*

Dr. Walker: Encourage the help of the public to protect dark skies. Make responsible use of outdoor lighting. Use shielded lighting, energy efficient lighting, and turn off lights (timers and motion sensors can help) when not needed.

Author: *If you could give just one piece of advice to students who are thinking about a career in astronomy, what would it be?*

Dr. Walker: Use your passion for the night sky to fuel your desire to excel in mathematics and physics.

What's a Star Party Really Like?

Just as in the case of any large gathering of people, at a star party you will find many different sorts of people, engaged in various activities. As we noted earlier, a star party gives you the opportunity to have an observing buddy or even to find a mentor to help with those more difficult skills. In one corner someone will make use of binoculars to find the Messier objects. In another area you may find a more experienced astronomer sharing views through the telescope with newcomers. Someone else may be pointing their 'scope at the numerous double stars and variable stars to be found. Another amateur may have his or her 'scope set up for astrophotography. You will find people using all kinds of different 'scopes—Schmidt-Cassegrains, small refractors, large reflectors, and perhaps even some extremely large Dobsonian reflectors aiming for dim targets. Most will leave before midnight, while a few aficionados may stay until dawn. Some people like to observe in groups, while others are more solitary. Certain amateur astronomers enjoy the quiet solitude that sometimes comes with observing while others are more gregarious. There is a personality type and observing preference for everyone (Fig. 10.2).

Most astronomy clubs offer one or two of these gatherings each month, one open to the public and perhaps the other for their members only. In most cases, monthly star parties are held at a location that has reasonably dark skies. Many clubs also provide loaner telescopes that can be used by members. The focal point of a star party is the observing, so a strict lights out policy is vital to ensure everyone can maintain their adaptation to the dark. Some clubs get everyone together to enjoy food a few hours before nightfall. Much of a star party becomes what you want to make of it. It can be a time for outreach, or a more quiet time for serious observing. Or it can be a more casual viewing session where you can share views of favorite objects while you enjoy the company of friends and family.

During the twentieth century large star parties of people from around the United States became popular. For example, during the 1930s hundreds of people would form long lines every Friday night for a chance to look through the famous 60-in.

Fig. 10.2 Image courtesy of IAU/A. Huggett/IYA2009

telescope on Mount Wilson. But gatherings of amateur astronomers who could build and observe with their own telescopes also began to emerge. Among the first of these to take place was the famous Stellafane Convention, which began in 1920. The first meeting was a gathering of 15 men and one woman who were all interested in learning how to make their own telescopes. By December 7, 1923, their meetings were officially held under the new club name: Springfield Telescope Makers, Inc. Ever since then, the annual Stellafane Convention has been known for its bringing amateur telescope makers together to exchange ideas, as well as to provide an excellent atmosphere for one of the largest star parties in the country.

Large star parties that sometimes include hundreds of people are now organized all over the United States. Some of these are the Enchanted Skies Star Party in New Mexico, the Texas Star Party, the Okie-Tex Star Party in Oklahoma, and the Orange Blossom Special Star Party in Florida to name just a few. In England, one of the largest astronomical gatherings is at the Autumn Equinox Sky Camp. This is a one-week-long event held in Kelling Heath, Norfolk.

These star parties go far beyond just a gathering of astronomy enthusiasts for a few hours. Since many people travel many hours to enjoy dark skies with fellow amateurs, comfortable overnight accommodations are available, plenty of food and activities. The activities can include games, tours, lectures, picnics, vendor displays, daytime entertainment and much more.

Most of all, these occasions allow for people to gather from many areas in order to share thoughts, ideas, and experiences related to astronomy. That has always been what astronomy is about, the passing of knowledge from one person to another, doing everything you can to expand on that knowledge, and then in turn using it to help someone else to grow. Sometimes it is just a matter of letting someone see your enthusiasm for astronomy. Often this is all that is needed to inspire someone else to get excited about the observable universe. So then, if you have not already done so, try out a star party, whether it's a local one or a large national one. You may be amazed at how much you learn from others. You may also be surprised at seeing just how much your own enthusiasm can rub off on other people.

Helpful Things to Consider

How can I make sure that I give an enjoyable presentation to my audience? Where can I find ideas for starting an astronomy club?

Why should the protection of our dark skies be an integral part of astronomy outreach? Where can I find a receptive audience for our astronomy outreach programs?

Reaching out to others so as to inspire them to be in awe of the vast universe that we now can see with the aid of telescopes is an effort that can be either reactive or proactive. Both ways of interaction are good. We need to be ready when called upon to provide the needed inspiration through various prepared programs and presentations. However, we also should seek out those groups of people that may be interested in astronomy or would be receptive to learning about science in general, and the impact that light pollution has on our planet. The more we use astronomical ideas and terms in our everyday communication, the more it may become prominent in the minds of everyone. Instead of losing the part of our culture that is inspired by dark skies, we can grow that part of our culture.

On the surface, progress in areas of outreach and light-pollution control can seem to be overwhelming for the individual. By working together each one of us can make a difference. Immediate and great change in these areas is not a reasonable expectation. There is a saying that says, "A yard is hard, but an inch is a cinch." Gradual and incremental change is a reasonable expectation. That change begins with us as individuals. If we keep making small strides in these important areas of amateur astronomy, eventually we can win back the heritage of our pristine dark skies. Remember, too, that no one is alone in the desire for these changes. There are many more organizations and programs outside of the few mentioned here that you can partner with to facilitate awareness in your local community. No one can inspire someone else without first feeling inspired. So what are you waiting for? Time to get out under the starry night sky to refuel your enthusiasm and awe for all that the universe has to offer.

Appendix:
The Greek Alphabet

The Greek alphabet is still used to designate the brighter stars of each constellation. This method of identification was developed by German astronomer Johan Bayer in 1603 when he published *Uranometria*, the first star atlas covering the entire sky.

α	Alpha	η	Eta	ν	Nu	τ	Tau
β	Beta	θ	Theta	ξ	Xi	υ	Upsilon
γ	Gamma	ι	Iota	o	Omicron	φ	Phi
δ	Delta	κ	Kappa	π	Pi	χ	Chi
ε	Epsilon	λ	Lambda	ρ	Rho	ψ	Psi
ζ	Zeta	μ	Mu	σ	Sigma	ω	Omega

© Springer International Publishing Switzerland 2015
D. A. Jenkins, *First Light and Beyond*, The Patrick Moore Practical Astronomy Series,
DOI 10.1007/978-3-319-18851-5

References[1]

1. DiCecco, A., Becucci, R., Bono, G., et al. (2010). On the absolute age of the Globular Cluster M92. *Publ Astron Soc Pacific, 122*(895), 991–999.
2. Hendry, M. A., Smartt, S. J., Crockett, R. M., et al. (2006). SN 2004A: another Type II-P supernova with a red supergiant progenitor. *Mon Not R Astron Soc, 369*(3), 1303–1320.
3. Fitzsimmons, A. (1993). CCD Stromgren UVBY photometry of the young clusters NGC 1893, NGC 457, Berkeley 94 and Bochum 1. *Astron Astrophys Suppl Ser, 99*(1), 15–29. ISSN 0365–0138.
4. Quinn, S. N., White, R. J., Latham, D. W., et al. (2012). Two "b"s in the beehive: the discovery of the first hot Jupiters in an open cluster. *Astrophys J Lett, 756*(2), Article ID. L33, 5 pages.
5. Kiss, L., Szabo, M., Balog, Z., et al. (2008). AAOmega radial velocities rule out current membership of the planetary nebula NGC 2438 in the open cluster M46. *Mon Not R Astron Soc, 391*(1), 399–404.
6. Crampton, D., Cowley, A., Schade, D., et al. (1985). The M31 globular cluster system. *Astrophys J, 288*, 494–513. Part 1 (ISSN 0004–637X).
7. Van Dyk, S., Schuyler, D., Zheng, W., et al. (2013). The progenitor of supernova 2011dh has vanished. *Astrophys J Lett, 772*(2), Article ID L32, 5 pages.
8. Sato, M., Hirota, T., Honma, M., et al. (2007). Absolute proper motions of water masers in NGC 281 measured with VERA, astrophysical masers and their environments. *Proc IAU Symp, 242*, 170–171.
9. Mundell, C., James, P., Loiseau, N., et al. (2004). The unusual tidal dwarf candidate in the merger system NGC 3227/3226: star formation in a tidal shock? *Astrophys J, 614*(2), 648–657.

---

[1] This research has made use of NASA's Astrophysics Data System.

© Springer International Publishing Switzerland 2015

D. A. Jenkins, *First Light and Beyond*, The Patrick Moore Practical Astronomy Series,
DOI 10.1007/978-3-319-18851-5

Index

© Springer International Publishing Switzerland 2015
D. A. Jenkins, *First Light and Beyond*, The Patrick Moore Practical Astronomy Series,
DOI 10.1007/978-3-319-18851-5